BINGE DRINKING IN ADOLESCENTS AND COLLEGE STUDENTS

Binge Drinking in Adolescents and College Students

Cecile A. Marczinski,
Estee C. Grant
and
Vincent J. Grant

Nova Science Publishers, Inc.
New York

For permission to use material from this book please contact us:
Telephone 631-231-7269; Fax 631-231-8175
Web Site: http://www.novapublishers.com

NOTICE TO THE READER

The Publisher has taken reasonable care in the preparation of this book, but makes no expressed or implied warranty of any kind and assumes no responsibility for any errors or omissions. No liability is assumed for incidental or consequential damages in connection with or arising out of information contained in this book. The Publisher shall not be liable for any special, consequential, or exemplary damages resulting, in whole or in part, from the readers' use of, or reliance upon, this material.

Independent verification should be sought for any data, advice or recommendations contained in this book. In addition, no responsibility is assumed by the publisher for any injury and/or damage to persons or property arising from any methods, products, instructions, ideas or otherwise contained in this publication.

This publication is designed to provide accurate and authoritative information with regard to the subject matter covered herein. It is sold with the clear understanding that the Publisher is not engaged in rendering legal or any other professional services. If legal or any other expert assistance is required, the services of a competent person should be sought. FROM A DECLARATION OF PARTICIPANTS JOINTLY ADOPTED BY A COMMITTEE OF THE AMERICAN BAR ASSOCIATION AND A COMMITTEE OF PUBLISHERS.

LIBRARY OF CONGRESS CATALOGING-IN-PUBLICATION DATA
Available upon request

ISBN: 978-1-60692-037-4

Published by Nova Science Publishers, Inc. ✦ *New York*

To Chris and Isabella
C.M.

To our families
E.G. and V.G.

CONTENTS

LIST OF FIGURES

LIST OF TABLES

Chapter 1

INTRODUCTION

In an article published in USA Today titled "Five binge-drinking deaths 'just the tip of the iceberg'", writer Robert Davis reported that the month of September 2004 had been one of the deadliest for binge-drinking college students. Five undergraduates in four states appeared to have drunk themselves to death, by accident, and police investigations had indicated that friends of the victims had sent their now deceased friends to bed thinking that they would sleep off their intoxication. Samantha Spady, 19, was found dead in a Colorado State University fraternity house. Lynn Gordon Bailey Jr., 18 was found dead at a University of Colorado fraternity house. Thomas Hauser, 23, was found dead in his apartment near Virginia Tech. Blake Hammontree, 19, was found dead in a fraternity house at the University of Oklahoma. Bradley Kemp, 20, was found dead at home at the University of Arkansas. All but one of the victims were underage. Police investigations further stated that these deaths by alcohol occurred because the person was too intoxicated to maintain his or her own airway. Death resulted when the person then suffocated on his or her own vomit or some otherwise harmless obstruction, such as a pillow or blanket (Davis, 2004).

National news stories such as the one described above really are the tip of the iceberg of the problem of binge drinking in adolescents and college students. Far too many parents have stayed up late worrying about their teenagers who are out with friends, who may be drinking themselves or end up in a car with someone who is too drunk to drive. Far too many parents send their child off to college only to receive the dreaded phone call that their child is in an emergency room, unconscious, vomiting and alone. Far too many parents have had to bury their children instead of watching them graduate from college because of drinking related accidents. We as a society have an uneasy relationship with alcohol. While many people drink responsibly, don't drive when intoxicated, and generally act in control of themselves when drinking, there are many others who don't. Those individuals who drink to get drunk and then get themselves into all kinds of trouble include, in large part, adolescents and college students.

The tragedy of the death of Samantha Spady is a telling example. She was a homecoming queen, class president, cheerleading captain and honor student in high school. Clearly beautiful, smart and charismatic, she attended college at Colorado State University. She died from binge drinking on September 5, 2004, because she drank between 30 to 40 beers and shots in 11 hours. Similar to many other college students, Samantha had been attending a big social event for the university where heavy drinking was omnipresent. It was the Colorado State University football game against the University of Colorado – Boulder. In the course of the celebrations and parties, Sam drank a great deal. So did others. She drank so much that she was left in an empty room in a fraternity house to sleep off her intoxication. A fraternity member found her dead while giving his mother a tour of the house. Her story has been retold in a documentary called *Death By Alcohol*.

Is her story tragic? Absolutely. Is her story unique? Unfortunately not. The National Institute of Alcohol Abuse and Alcoholism (NIAAA) estimates that approximately 1,400 college students ages 18 to 24 die each year from alcohol-related incidents. For this reason, binge drinking has been argued to be the number one public health hazard and the primary source of preventable morbidity and mortality for the more than 6 million college students in the United States (Wechsler et al., 1995). Epidemiological evidence has shown that binge drinking is widespread on college campuses, with almost half of the students reporting binge drinking. In addition, binge drinking has been associated with unplanned and unsafe sexual activity, assaults, falls, injuries, criminal violations, and automobile accidents (Wechsler et al., 1995, 1998, 2000). In the United States, approximately 500,000 college students are injured and 1,700 die each year from alcohol-related injuries (Hingson et al., 2005).

Is college student binge drinking a problem limited to college campuses? The answer is a resounding "No." Recent studies present compelling evidence that many of the problems with heavy drinking evident on college campuses continue a pattern initiated in high school (Miller et al., 2007). Underage drinking is the leading public health and social problem in the United States and is associated with the three leading causes of death among teenagers: unintentional injuries, homicides, and suicides. Underage drinking leads to numerous harmful health and societal consequences. Despite significant declines in tobacco and illicit drug use among teens, underage drinking has remained at consistently high levels. Teenagers who drink tend to neglect responsibilities, get into fights, miss school, drive after drinking, and engage in suicidal behavior. Furthermore, underage drinking is associated with using illicit drugs, having unprotected sexual activity and carrying weapons (Miller et al., 2007). Moreover, underage drinking has become the domain of even younger children. A recent U.S. survey found that over 11% of 12 year old children (6th graders) say they have used alcohol. By age 13, that number doubles. By age 14 (8th graders), 41% of children state that they have had at least one drink, and nearly 20% say they have been drunk at least once (U.S. Department of Health and Human Services, 2007).

The problem of underage drinking is finally starting to get the attention that is due. The 2005 National Survey on Drug Use and Health estimated that there are 11 million underage drinkers in the United States and 7.2 million of them are considered binge drinkers. In its first "Call to Action" against underage drinking, the U.S. Surgeon General's Office appealed to Americans to do more to stop the 11 million current underage drinkers in the United States from using alcohol, and to keep other young people from starting (U.S. Department of Health and Human Services, 2007). As part of a news release, Acting Surgeon General Kenneth Moritsugu, M.D., M.P.H. stated that, "Too many Americans consider underage drinking a rite of passage to adulthood. Research shows that young people who start drinking before the age of 15 are five times more likely to have alcohol-related problems later in life. New research also indicates that alcohol may harm the developing adolescent brain. The availability of this research provides more reasons than ever before for parents and other adults to protect the health and safety of our nation's children."

If we are to answer the call of the U.S. Surgeon General to reduce binge drinking, we first need to have a comprehensive look at the scope of this problem. The purpose of this book is to examine all aspects of binge drinking in adolescents and college students. Our goal in writing this book was to provide the most comprehensive and recent information about the harms and hazards associated with this type of alcohol consumption. In the past few years, there has been a dramatic rise in the amount of scientific information about binge drinking that has yet to be compiled into one resource. As such, there is now a need for a comprehensive resource that contains all of the newest information about binge drinking and how this type of alcohol consumption (drinking to get drunk) differs from social drinking and chronic alcohol dependence. Furthermore, recent successful intervention techniques both on the individual level (e.g., motivational interviewing in the emergency room) to the institutional level (e.g., programs to reduce binge drinking on college campuses) are discussed.

This book covers the most recent research on binge drinking and its causes and consequences. Chapter 2 provides the details of how and why practitioners and researchers distinguish binge drinking (i.e., drinking to get drunk) from social drinking and chronic drinking/alcoholism. Chapter 3 addresses the scope of the binge drinking problem by summarizing the findings from epidemiological studies and presents demographic trends both in the United States and other countries around the world. Chapter 4 discusses a serious problem for binge drinking research in that investigators have difficulty agreeing on what constitutes a binge and who should be called a binge drinker. Chapter 5 examines how binge drinking is a social phenomenon and how peer norms, fraternity/sorority membership, college athletics, and current alcohol policies are environmental variables of importance in the binge drinking problem.

Chapter 6 examines the neurocognitive and health effects that binge drinking has on young people. Binge drinking is often episodic, such as when a high school student gets drunk on Saturday night but then doesn't drink for the rest of the week. Episodic binge drinking doesn't appear to offer any protection to the brain, however. A variety of studies have now observed that young adult binge drinkers suffer neurocognitive deficits, perhaps because intoxicating doses of alcohol are damaging a brain that is still in development. Chapter 7 reports on animal studies that supports this claim. The binge consumption of alcohol can cause brain damage, particularly in the prefrontal cortex and amygdala regions. While the exact mechanisms of why binge alcohol use causes brain damage still need to be elucidated, it is clear from these carefully controlled animal studies that this pattern of drinking causes serious neurological harm.

Chapters 8 and 9 examine the risky behaviors undertaken by binge drinkers that lead to various harms and hazards. In Chapter 8, the relationship between binge drinking and impaired driving is discussed. Binge drinkers are the demographic group that is most likely to drive while impaired (despite popular notions that all impaired drivers are alcoholics). Binge drinkers also take other risks such as riding with an intoxicated driver and not wearing a seatbelt. Binge drinkers take additional health risks, such as engaging in risky sexual behaviors. Chapter 9 examines the common consequence of binge drinking, a visit to the Emergency Department. Alcohol related visits to the Emergency Department are far too common among high school and college students. These visits stem from two main reasons: acute alcohol intoxication, and injury or trauma. Heavy alcohol consumption impacts the likelihood of injury, as well as the type and severity of injury. Practitioners have realized that these Emergency Department visits are an important opportunity to intervene in reducing hazardous drinking behavior. Chapter 10 details how these patients can be screened appropriately and how brief interventions can be provided within the context of an Emergency Department. Adolescents and college students who present to the Emergency Department are potentially in an enhanced state of receptiveness to intervention and readiness to change their drinking behavior. This is an ideal opportunity to potentially reduce future hazardous drinking and its associated risk of injury, illness, and even death.

Chapters 11 through 16 discuss reasons why binge drinking occurs and why young people drink to such extremes. Chapter 11 discusses the relationship between lack of self-control and binge drinking. When an individual engages in binge drinking, control over thought and action seems to be lost as the individual continues to keep drinking beyond the threshold of what is safe to the individual and to others around the individual. This chapter discusses what is known about self-control and why self-control appears to get hijacked when drinking alcohol. Various laboratory studies have provided compelling evidence that alcohol directly impairs self-control and that diminished self-control under alcohol is significantly worse for binge drinkers. Chapter 12 demonstrates how poor self-control can have widespread effects on a variety of behaviors, including heightened aggression, inappropriate sexual behavior and a further escalation of drinking. There may be a vicious cycle at play between failures of self control and alcohol consumption contributing to binge drinking. It may be that only a small amount of alcohol is required to decrease self control, which leads to further drinking. Unfortunately, binge drinkers appear to be completely unaware of this process. When drinking, they have different subjective reactions to alcohol, including feeling less intoxicated, compared to their moderate drinking peers. This reduced intoxication level may explain the high rates of impairing driving for binge drinkers. The binge drinkers feel less impaired by alcohol and are more willing to drive even though their actual driving behavior remains poor. Chapter 13 looks at the role of alcohol tolerance in understanding binge drinking. As drinking increases in frequency, tolerance to the effects of alcohol occurs (i.e., the person feels less sensitive to alcohol effects). There are several types of tolerance, including acute tolerance which occurs over the course of a drinking session. Developing acute tolerance for the subjective effects of alcohol may set up a significant safety risk for binge drinkers. As blood alcohol declines, binge drinkers may feel less intoxicated and think that they are capable to drive home. However, a variety of motor skills remain impaired, including driving.

Chapter 14 looks at who is most at risk for binge drinking, even before the first drink is ever consumed. A variety of risk factors have now been identified, including personality traits such as impulsivity and sensation seeking. Some biological and social risk factors have also been identified for binge drinkers. Chapter 15 examines the role that beliefs and expectations about alcohol play in driving binge drinking behavior. Cognitive factors, such as the role of expectation, can exert extremely powerful influences on alcohol use. Alcohol expectancies can be positive (e.g., alcohol makes people more sociable) or negative (e.g., alcohol makes people act inappropriately). There is an important developmental change whereby young children often hold negative expectancies about alcohol use whereas older children hold more positive expectancies. Since positive expectancies are associated with the onset of drinking, any effort to delay this shift from negative to positive expectancies should delay the onset of binge drinking in adolescents. Expectancy challenge interventions appear successful in changing alcohol expectancies and may help moderate binge drinking in both adolescents and college students.

Chapter 16 discusses the recent trend in mixing alcohol with energy drinks. While not widely studied yet, limited evidence suggests that alcohol mixed with energy drinks (AmED) is extremely popular in high school and college students and that AmED consumption may lead to riskier behavior than alcohol consumption alone, including an escalation in drinking. Users of AmED frequently report greater pleasure and less sleepiness when using AmED, compared with alcohol alone. Reduced feelings of sedation following drinking may lead to further drinking and increase the likelihood that a binge will occur. Chapter 17 discusses the variety of prevention and intervention strategies that have been used to try to reduce binge drinking at the elementary school, high school and college level. The DARE program and social norms programs have had limited success in reducing binge drinking while skills training programs have fared better in evaluations (but have not been widely implemented). New ideas and efforts on prevention and interventions with the problem of binge drinking in adolescents and young adults are clearly still needed. Finally, Chapter 18 summarizes key findings and what we still need to learn to better understand binge drinking in young people.

This book discusses the psychological, medical and societal implications of the rise in binge drinking in adolescents and college students. Where possible, we have tried to discuss the U.S. binge drinking problem with reference to similarities and differences with other countries around the world. While binge drinking in young people is of significant concern in many countries around the world, rates of binge drinking in the U.S. are much higher compared to most other countries (with the exception of the United Kingdom, a country that also suffers from seriously high rates of binge drinking). The evidence from both human and animal research has demonstrated that adolescents respond differently to alcohol than adults, and the young adolescent brain appears to have differential sensitivity to alcohol-induced brain damage compared with adults. Psychologists, physicians, neuroscientists, nurses, social workers, injury preventions professionals, school counselors, and high school teachers will all benefit from the material presented in the following chapters. By disregarding binge drinking in adolescents and college students, we do these young people a great disservice. If we evaluate the evidence that binge drinking is harmful to health and well-being, we may be in a better position to teach our young people how to change their behavior. Their future is bright; engaging in binge drinking dims their prospects on so many levels. It is our obligation to do what we can to help these young people achieve their goals and realize their potential.

Chapter 2

DIFFERENTIATING BINGE DRINKING FROM SOCIAL DRINKING AND CHRONIC DRINKING/ALCOHOLISM

Why do we need to distinguish binge drinking from other alcohol consumption patterns such as social drinking and alcohol abuse and dependence? In the past, binge drinking was never thought of as a problem, and was simply seen as part of the continuum of social drinking, albeit to the extreme. Even today, there are many individuals in the general public who believe that incidences of getting drunk are part of celebrations or part of a developmental stage. For example, a teenager gets caught getting drunk and is grounded by his or her parents for this infraction, but the parents think this is part of growing up. The teenager then ages and attends college where he or she attends parties and gets drunk. Upon graduation, however, the parents assume that their child should 'mature' out of this alcohol consumption pattern because the demands of work and other aspects of adult life prevent such frivolity of youth.

College life seems to have special status in the notion that binge drinking is developmentally normal. College students have always had higher rates of drinking than their peers who do not attend college. Back in the 1950s, Straus and Bacon provided data from a national survey documenting the widespread use of alcohol by college students (Straus and Bacon, 1953). Thus, college drinking has long been considered a rite of passage. To a certain extent, the stereotype of the white, male, athletic member of a fraternity who views parties as important to the college experience, and also is a heavy college drinker, has been found to be true. Similarly, the notion that these drinkers are not passive victims of peer pressure, but instead are willing participants of this type of alcohol consumption has also been observed to be true (Wechsler et al., 1995). However, the rise in college binge drinking to rates as high as almost one half of all college students and its associated serious health consequences has demonstrated that, far from being an innocent rite of passage, college binge drinking is a serious and sometimes deadly problem. Research has shown that binge drinking consists of a drinking style characterized by more frequent and heavier alcohol use, intoxication, and drinking with the purpose to get drunk (Wechsler et al., 1994, 1995). The rise in binge drinking has in part occurred against an overall general decline in drinking. There has been an overall decline in drinking alcohol in U.S. society, yet binge drinking on U.S. college campuses has risen (Johnston et al., 1991). Furthermore, the rise in binge drinking is now extending to a demographic of even greater concern. Binge drinking is now a pervasive problem in high school students, with underage drinking contributing to the three leading causes of death among adolescents (Miller et al., 2007). The goal of this chapter is to understand why binge drinking should be considered separately from responsible social drinking, and yet does not fall in domain of the medical illness of alcoholism.

SOCIAL DRINKING

When one asks college students if they are social drinkers, most would answer in the affirmative. However, social drinking differs from binge drinking. Social drinkers typically have one to three drinks during an occasion and then stop (Fillmore, 2001; Maisto et al., 2008). Social drinking is thought to be associated with no ill effects and in fact may be associated with various health benefits in adults. While heavy drinking is associated with various deleterious health consequences and may spiral into alcohol abuse and dependence, a number of studies have noted that moderate social drinkers are actually healthier than individuals who completely abstain from alcohol consumption. Research on the relationship between moderate alcohol consumption and health benefits originated from the French paradox. Epidemiologists found that individuals living in France consumed a diet high in saturated fats and yet had a low incidence of coronary heart disease (Frunkel et al., 1993). This biological paradox was originally thought to be due to high consumption of red wine in the French, with the consumption of one to two 4-ounce glasses of wine a day being associated with improved health effects (Renaud and de Lorgerd, 1992). The benefits to cardiovascular health may not be specifically limited to wine but instead may be due to the effect of any kind of alcohol (Gronbaek et al., 1995). This matter remains in debate as some have argued that components of red wine, in particular, may be beneficial to vascular health (Corder et al., 2006).

For the most part, studies of the health benefits of moderate alcohol consumption have focused on two major factors: the risk of cardiovascular disease and mortality. Moderate drinking is associated with lower risk of coronary disease (Marmot, 2001). Individuals who consume between one to three drinks a day have been found to have increased production of high density lipoproteins (HDLs), which removes damaging cholesterol from artery walls (Gaziano et al., 1993; Rimm, 2000). The amount of alcohol required to elicit health benefits does not appear to be large. For example, a longitudinal study from Denmark found that the heart health benefits of the moderate consumption of alcohol was evident in men who had between 7 and 35 drinks a week and in women who had as little as one drink a week (Tolstrup et al., 2006). Similarly, the moderate consumption of alcohol has been shown to be associated with greater longevity. For example, a prospective study of Chinese men found that consumption of no more than two drinks per day was associated with a 19% reduction in mortality risk. The authors found this protective effect for all types of alcoholic drinks consumed (de Groot and Zock, 1998).

Alcohol is also used sometimes for therapeutic reasons. For example, alcoholic beverages are sometimes recommended in moderate amounts to elderly patients for stimulation of appetite and digestion before meals (Maisto et al., 2008). Alcohol reduces the amount of fat in the body that is oxidized. This acute effect of alcohol accumulates to result in long-term increased body fat and weight gain when alcohol is used in addition to normal food intake (Suter et al., 1992). For a frail underweight elderly person with a poor appetite, weight gain improves health. However, weight gain associated with alcohol may contribute to health problems for other social drinkers, especially in American society where obesity is a significant concern.

Thus far, research on the health benefits of moderate alcohol consumption has been limited in scope and largely has focused on cardiovascular function. It is still unclear how moderate alcohol consumption impacts other aspects of health, including liver function, cancer risk or cognitive functioning. Furthermore, social drinking may not be appropriate for all individuals or certain subgroups in particular, such as the elderly who take certain medications and/or individuals with coexisting illnesses (Moore et al., 2006; Saitz, 2005).

There are good reasons why underage drinking, even in a moderate fashion, should be avoided. First, social convention and legal restrictions prohibit it. In most countries, and in particular in the United States and Canada, underage drinking is illegal. Second and more importantly, the brain continues to develop well into adolescence, and consumption of alcohol may impact this development. Adolescent neurodevelopment occurs in brain regions associated with motivation, impulsivity, and addiction. Maturational changes in the frontal cortical and subcortical monoaminergic systems confer greater vulnerability to the addictions of alcohol and drugs, especially for adolescents. Therefore, moderate consumption of alcohol may be fine for an adult brain that is finished development but inappropriate and dangerous for an adolescent brain that is still its last phase of neurodevelopment (Chambers et al., 2003).

BINGE DRINKING

Binge drinkers do not limit their consumption of alcohol to a few drinks during an occasion (social drinking) but instead continue to drink, often to the point of intoxication (getting drunk). This is what distinguishes social drinking from binge drinking. Binge drinking is commonly defined as consumption of five or more drinks during a single occasion, a pattern of drinking that is particularly prevalent among high school and college students (Presley et al., 1995; Wechsler et al., 1994). Binge drinking is associated with a host of social and personal problems. Binge drinkers are more likely to display poor academic performance, drive while intoxicated, damage property, suffer injuries, and engage in violence and risky sexual behavior (Wechsler et al., 1994). A continued pattern of binge drinking also poses immediate and long-term health consequences. The immediate consequences of this type of alcohol consumption include health risks such as alcohol poisoning and alcoholic hepatitis. In the long-term, the health consequences of this type of alcohol consumption includes alcohol dependence and liver cirrhosis (Wechsler et al., 1995).

Since binge drinking is associated with substantially increased risks of acute health problems, binge drinking is now argued to be the number one public health hazard and the primary source of preventable morbidity and mortality for the more than 6 million college students in the United States (Wechsler et al., 1995). However, binge drinkers themselves generally do not seem to have concern about their alcohol consumption. In surveys, most binge drinkers do not consider themselves to be problem drinkers and have not sought any treatment for an alcohol-related problem. Similarly, alcohol being illegal appears to be of little importance to most undergraduate college students in the U.S., as alcohol continues to be used widely on most college campuses (Wechsler et al., 1994). The same observations regarding perceptions of underage drinking have been made in adolescent binge drinkers, with these individuals appearing equally unconcerned about the health risks and legal ramifications of their behavior (Miller et al., 2007).

Binge drinking differs from social drinking mainly in the amount of alcohol consumed within one episode. While a social drinker may consume one drink daily over the course of the week, a binge drinker may consume seven drinks in one evening and then not consume any alcohol for the remaining six days of the week. Even though the amount of alcohol is the same, the consumption of all of it in one episode creates a problem. Alcohol acts on the body as a depressant, and its acute effects are proportional to the magnitude of the blood alcohol concentration (BAC), a measure of the amount of alcohol in the bloodstream. Breath, blood and urine samples are all ways to measure BAC. For example, BAC is commonly measured by police departments by breath sample because of the known ratio (1:2,100) between the amount of alcohol in the lungs and the amount in blood. Since gas chromatography provides excellent estimates of BAC, they are considered legally admissible evidence (Maisto et al., 2008).

Table 2-1 illustrates the effects associated with various blood alcohol concentrations (BACs). Since BAC measures the amount of alcohol in the bloodstream, a BAC of 0.10g% means that a person has one part of alcohol per one thousand parts of blood (Maisto et al., 2008). Note that the effects associated with various BACs will depend on various factors since a person's BAC level is affected by many variables.

Table 2.1. Typical acute effects of alcohol associated with different ascending blood alcohol concentrations (modified from Maisto et al., 2008 and Wechsler and Wuethrich, 2002). Note that a standard drink refers a half an ounce of alcohol (e.g., one 12-oz. beer, one 5-oz. of glass of table wine, or one 1.5-oz. shot of distilled spirits) (NIAAA, 2004).

BAC (g%)	Quantity Consumed in about 1 Hour		Effects
	160 pound male	120 pound female	
0.01-0.02	1	0.5	Slight changes in feeling; sense of warmth and well-being; a driver's ability to divide attention between two or more tasks (dialing a cell phone while driving) can be impaired by BACs of .02% or lower.
0.03-0.04	1	0.5	Feelings of relaxation, slight exhilaration, happiness; skin may flush; mild impairment of motor skills.
0.05-0.06	2	1	Legal level of intoxication is .05% in many European countries. Effects become more noticeable. Feelings of warmth, relaxation and mild sedation. More exaggerated changes in emotion, impaired judgment, and lowered inhibitions. Coordination may be altered. Visual and hearing acuity decreased. Slight speech impairment. Mild working memory impairment (e.g., forgetting someone's name after they've been introduced to you).

BAC (g%)	Quantity Consumed in about 1 Hour		Effects
	160 pound male	120 pound female	
0.07-0.09	3	2	Legal level of intoxication is .08% in all U.S. states and Canadian provinces. Noticeable speech impairment and disturbance of balance; impaired coordination; major increase in reaction time; may not recognize impairment. Sensory feelings of numbness in cheeks, lips, and extremities. Increased urination. Further impairment of judgment.
0.10-0.13	4	3	Noticeable disturbance of balance (e.g., person may stagger); uncoordinated behavior; significant increase in reaction time; increased impairment of judgment and memory.
0.14-0.17	6-7	4	Significant impairment of all physical and mental functions; difficulty in standing (balance) and talking. Large increase in reaction time. Large impairment in judgment and perception; cannot recognize impairment; blackouts (failure to recall events for a portion of time) are observed at BAC levels about .15%
0.20-0.25	9	5	Difficulty staying awake; confused or dazed; substantial reduction in motor and sensory capabilities; slurred speech, double vision, difficulty standing or walking without assistance.
0.30-0.35	10	8	Confusion and stupor. Difficulty comprehending what is going on; general suspension of cognitive abilities; possible loss of consciousness (passing out); vital reflexes like gagging and breathing become compromised. Lethal dose (LD) 1 in humans (the level at which 1 out of 100 people would die at this BAC).
0.40	15.5	10	Typically unconsciousness/coma; sweatiness and clamminess of the skin. Alcohol has become an anesthetic.
0.45-0.50+	16+	10+	Deep coma/death. Circulatory and respiratory functions may become totally depressed. Lethal dose (LD) 50 in humans (the level at which 50 out of 100 people would die at this BAC) has been estimated to be reached after the consumption of 25 standard drinks in an hour.

How much an individual drinks and how quickly, gender, body weight, percentage of body fat (physical condition), drugs and medications being used, and the amount of food in the stomach are all factors that influence BAC. Changes in body and brain functioning as BAC rises occur because the alcohol alters activity in various body systems. For example, at very low BACs starting at .02g%, increased dopamine levels in the nucleus accumbens area of the brain elicits the pleasurable feelings produced by alcohol. At BACs around .05g%, activity in the brain's cerebellum results in loss of motor coordination which worsens as BAC rises. At BACs around .06g%, activity of the brain's frontal lobes becomes altered resulting in working memory impairments, such as forgetting someone's name after an introduction. As BAC rises to .08g%, the antidiuretic hormone normally released by the hypothalamus resulting is suppressed, which results in increased urination.

The medulla is a brain structure located above the spinal cord at the base of the brain, which regulates basic life functions, such as breathing, heart rate, vomiting, swallowing, and blood pressure. When toxic chemicals reach high levels in this area, the vomit center is triggered in an adaptive attempt to purge the body of further toxins, which is why drinking large quantities of alcohol often causes nausea and vomiting. If vomiting cannot occur, circulatory and respiratory functions can become totally depressed and death by alcohol intoxication may result.

The recent rise in binge drinking in adolescents and college students may not be an entirely isolated phenomenon. Reports have indicated that binge drinking may be increasing among all age groups in the United States. Naimi et al. (2003) analyzed findings from the 1993-2001 surveys of the Behavioral Risk Factor Surveillance System, which are random-digit telephone surveys of adults ages 18 and older conducted annually in all states in the U.S. The annual sample sizes ranged from 102,263 in 1993 to 212,510 in 2001. The authors found that the total number of binge drinking episodes among U.S. adults increased from approximately 1.2 billion to 1.5 billion. This measured as an increase of 17% in binge drinking episodes per person per year over the time period. This increase appeared to be escalating over time as the binge drinking episodes per person per year increased by 35% between 1995 and 2001. Men accounted for 81% of binge drinking episodes (a gender difference more dramatic than in adolescents and college students where the gender gap is narrower or almost absent in some studies). While rates of binge drinking were highest among the 18 to 25 age group, 69% of binge drinking occurred among those individuals ages 26 and older. Delineating how binge drinking differs for alcohol abuse and dependence, Naimi et al. (2003) found that 47% of binge drinking episodes occurred among otherwise moderate (non-heavy drinkers) and 73% of all binge drinkers were moderate drinkers. Thus, U.S. per capita binge drinking episodes have increased since 1995 despite the fact that U.S. consumption of alcohol has been decreasing.

Some of the health effects and associated problems associated with intoxication will be discussed in other chapters in this book. However, the BAC chart in Figure 2-1 illustrates the numerous changes that occur in an individual as blood alcohol rises. Still, the effects and problems of intoxication are not limited to those listed in the BAC chart. For example, when high BACs are reached rapidly, a blackout may result. A blackout is a failure to recall events that occurred while drinking even though there was no loss of consciousness (Maisto et al., 2008). For example, a person may have no memory for events from the prior evening between midnight and two a.m. However, his friends may describe all of the things this person was doing during this time period. This amnesia is thought to result from a failure to transfer information from short-term memory to long-term memory (Maisto et al., 2008). The risks associated with blackouts increase if other dangerous behaviors were engaged in during this time period. For example, the individual may have no recall of taking other illegal drugs or engaging in risky sexual activities for the time period during the blackout.

ABUSE AND DEPENDENCE

The problems of intoxication and its associated health risks are also present with alcohol abuse and dependence. However, with chronic and abusive drinking, an individual becomes physically and psychologically dependent on alcohol. For the most part, this dependence is absent in binge drinkers. A college student may drink excessively on the weekend yet have no want for a drink during the week. This is not the case for the chronic and abusive drinker.

Since the 1950s, the American Psychiatric Association's (APA) formal diagnostic system is often used in the determination of whether an individual suffers from alcohol dependence. Its *Diagnostic and Statistical Manual* (DSM) provides the APA's formal diagnostic criteria for different mental illnesses or disorders. The updated version, DSM-IV-TR was published in 2000, and practitioners rely on the section called "substance-related" (alcohol- or other drug-related) disorders when assessing patients who might have a serious problem with alcohol. There are two diagnoses in this section: Substance Dependence and Substance Abuse (American Psychiatric Association, 1994, 2000).

Substance Dependence refers to a maladaptive pattern of substance use leading to clinically significant impairment or distress. To make the diagnosis, the practitioner must confirm the occurrence of three or more of the following at any time in the last year: 1) tolerance (i.e., use of more alcohol to achieve intoxication/effect or diminished effect with continued use of the same amount), 2) withdrawal (the characteristic syndrome when the substance is no longer used), 3) escalation of use, 4) persistent desire or unsuccessful efforts to cut down, 5) great deal of time spent in activities necessary to obtain or use or recover from use, 6) important social, occupational or recreational activities given up because of use, and 7) continued use despite knowledge of psychological or physical problems it is causing (American Psychiatric Association, 1994, 2000).

If drinking is causing serious problems but the individual does not meet the criteria for Substance Dependence, then the individual might fit the criteria for Substance Abuse. Substance Abuse also refers to a maladaptive pattern of substance use that leads to clinically significant impairment or distress. However, only one of the following features must be present to make this diagnosis: 1) use is resulting in failure to fulfill major role obligations at work, school or home, 2) use is becoming physically hazardous (e.g., impaired driving), 3) recurrent legal problems with use (e.g., arrests of public intoxication), or 4) continued use despite social or interpersonal problems (e.g., physical fights) (American Psychiatric Association, 1994, 2000).

For Substance Dependence (the more serious of the two diagnoses), the practitioner will investigate for evidence of physiological dependence. This is the case if the person exhibits tolerance or withdrawal. Tolerance refers to the increased amount of drug needed to achieve intoxication or the diminished drug effect with continued use of the same amount. Withdrawal refers to the definable illness that occurs with a cessation or decrease in use of the drug. Tolerance is discussed in greater detail in chapter 14.

As a brief overview, tolerance involves different mechanisms. Regular use of alcohol can result in an increase in metabolism of the drug so that the user can consume more alcohol. Frequent drinking also allows the brain and other parts of the central nervous system to become less sensitive to effects of alcohol. Tolerance can even have a learned component (e.g., you walk more slowly when drunk, since you know that alcohol makes you clumsy)

(Maisto et al., 2008). Tolerance can be dramatic in individuals who are severely alcohol dependent. As an example, arrests for impaired driving have been made when individuals had BACs so high that unconsciousness would normally be the case for a person who doesn't normally drink.

The development of withdrawal symptoms in alcohol dependence illustrates how much the person's body has become used to alcohol being present at all times, with symptoms of this physical dependence being quite severe in some cases. There are distinct phases of alcohol withdrawal syndrome. In Phase 1 of alcohol withdrawal, the individual may experience symptoms as soon as a few hours after drinking has stopped. Such symptoms can be present even if the blood alcohol concentration is still above zero. Symptoms include tremors (shakes), profuse sweating, and weakness. An intense need to seek alcohol will dominate consciousness. Other symptoms at this stage can include agitation, headache, anorexia (no appetite), nausea, vomiting, abdominal cramping, high heart rate, and exaggerated and rapid reflexes. In some cases, visual and auditory hallucinations (seeing things or hearing things that are not physically present, such as hearing voices) may occur.

Within 24 hours of drinking cessation, Phase 2 of the alcohol withdrawal syndrome occurs. The significant change in this phase is the possibility of developing grand mal seizures. The frequency of these seizures ranges from one isolated seizure to continuous seizure activity with little or no interruption. Phase 3 of the alcohol withdrawal syndrome commences at approximately 30 hours after drinking cessation. Phase 3 can last up to 3 or 4 days and commonly includes the development of delirium tremens (DTs). In this phase, severe agitation develops and the person may appear confused and disoriented. Nearly continuous activity is present as the person cannot stay still. A very high body temperature and abnormally rapid heartbeat are also present. The person may experience terrifying hallucinations at this point (in visual, auditory or tactile form). For example, tactile hallucinations may include feelings that bugs or little animals are crawling on the skin. Hallucinations can be accompanied by delusions (irrational thoughts). There is a high potential of violent behavior and medical management is often needed. Phase 3 can be quite dangerous as deaths during DTs occur due to the high fever, cardiovascular collapse or traumatic injury. By about 5 to 7 days after drinking stopped, the withdrawal process ends. The individual will be exhausted and severely dehydrated by this end point (Jacobs and Fehr, 1987).

For most adolescent and college age binge drinkers, the above criteria are not met. While some binge drinkers may experience problems with their drinking, many just do not fit the APA criteria for Alcohol Abuse or Alcohol Dependence. Despite the pattern of drinking associated with binge drinking being problematic, it is often unrecognized as a serious concern because the individuals in question do not meet the criteria for the medical illness of alcohol dependence.

BINGE DRINKING, ALCOHOL ABUSE AND DENIAL

There are no clear demarcations between social drinking and binge drinking, or between binge drinking and alcohol abuse and dependence. Many social drinkers may drink more alcohol once in a while (e.g., at a wedding) and many binge drinkers start tipping over into alcohol abuse and dependence (e.g., an incident of impaired driving or an arrest by campus police for public intoxication). However, there is one important similarity between binge drinkers and alcoholics. It has been observed that frequent binge drinkers (individuals who report having binged at least 3 times in a prior 2 week period) are similar to the alcohol abusers in their tendency to deny that they have a problem. Frequent binge drinkers experience numerous problems associated with their alcohol consumption. Because of their alcohol use, they are 10 times more likely to get into trouble with campus police, damage property or get injured compared to non-binge drinkers. When asked if they had driven a car after having five or more drinks, 40% of male binge drinkers admitted to having done so, compared to only 2% of male non-binge drinkers (for females, the rates were 21% for frequent binge drinkers versus 1% of non-binge drinkers). Despite all of these problems associated with their drinking, only 0.6% of frequent binge drinkers designated themselves as problem drinkers and a negligible few ever sought treatment for a problem with alcohol (Wechsler et al., 1994).

CONCLUSION

Binge drinking differs from other types of alcohol consumption, including social drinking and clinically diagnosed alcohol abuse and dependence. While social drinkers limit consumption to a few drinks per episode, binge drinkers often drink to a level of intoxication. However, binge drinkers do not routinely fit the criteria for alcohol abuse and dependence. They also lack the symptoms of tolerance and withdrawal which are indicative of physiological and psychological dependence on alcohol. Because of these important differences, binge drinkers should be considered separately from social drinkers and alcoholics, with unique concerns needing to be addressed.

Chapter 3

BINGE DRINKING DEMOGRAPHICS

To address the concern of binge drinking in adolescents and college students, we first need to examine the scope of the problem. This chapter covers recent epidemiological studies of binge drinking that have asked several important questions. For example, how many adolescents and college students binge drink regularly? Who is most likely to engage in this behavior? Who is least likely to engage in this behavior? Are teenagers and young people from other countries as likely to engage in binge drinking as their U.S. counterparts? What are the consequences of binge drinking to health and well-being? The answers to these questions provide the necessary background to understand why this problem has become so widespread and why we failed to stem the rise in binge drinking in adolescents and college students in the past few decades.

BINGE DRINKING IN ADOLESCENTS

Underage drinking and binge drinking in adolescents is widespread and known to cause significant health complications. In adolescence, rates of alcohol use start as low as 3.9% among teenagers aged 12 to 13 and then increase rapidly with age with rates as high as 51.6% among those aged 18 to 20 (Substance Abuse and Mental Health Services Administration, 2007). According to data from the "Monitoring the Future" study, 17.2% of eighth graders, 33.8% of tenth graders, and 45.3% of twelfth graders reported current alcohol use, defined as the consumption of at least 1 alcoholic beverage in the past 30 days (Johnston et al., 2007). Underage drinking is slightly more common in males than females (SAMHSA, 2007). These statistics are startling considering the fact that it is illegal in all states for these young people to be consuming any alcohol.

There is a paucity of population-based studies of binge drinking in youth. However, one study by Miller et al. (2007) analyzed data on current drinking, binge drinking and associated health risk behaviors from the 2003 National Youth Risk Behavior Survey (YRBS). This survey was developed by the Centers for Disease Control and Prevention and is a school-based survey that addresses drinking but also other health risk behaviors. The questionnaire queries six domains of health risk behaviors: 1) behaviors that lead to unintentional injury and violence, 2) tobacco use, 3) alcohol and other drug use, 4) sexual behaviors contributing to unintended pregnancy and sexually transmitted diseases, 5) unhealthy dietary behaviors, and 6) physical inactivity. The 2003 survey was administered to a nationally representative sample of approximately 15,000 students in grades 9 through 12 in both private and public schools across the United States.

When students were asked the question regarding any alcohol use (the question posed was "During the past 30 days, on how many days did you have at least 1 drink of alcohol?"), 45% of high school students reported drinking alcohol. When asked about binge drinking ("During the past 30 days, on how many days did you have 5 or more drinks of alcohol in a row, that is, within a couple of hours?"), 29% of high school students reported binge drinking. Note that this definition does not make a correction for gender, even though girls often weigh less than boys, and this definition does not correct for the overall smaller body weight of teenagers compared with adults. While girls reported more current drinking with no binge drinking, the binge-drinking rates were similar among girls and boys. Rates of binge drinking also increased with age and school grade.

The Miller et al. (2007) study received a fair amount of press coverage as the reported rates of binge drinking in high school students (almost 1/3 of high school students reported regular binge drinking) were much higher than those reported in other older studies. One reason for discrepancies in rates of binge drinking in high school students is that definitions of what we should call binge drinking differ dramatically among studies. The definition of binge drinking in the Miller et al. (2007) study (i.e., 5 or more drinks in a row in the past 30 days) was more similar to surveys that are given to college students.

By contrast, the Substance Abuse and Mental Health Services Administration (SAMHSA) also collects extensive data on rates of underage drinking as part of the National Surveys on Drug Use and Health. However, the SAMHSA definition of binge drinking differs dramatically as the focus of this survey is to identify individuals who are developing or have developed alcohol abuse problems. This survey is conducted annually by the Office of Applied Studies within SAMHSA. The survey includes households in all 50 U.S. states and the District of Columbia and includes individuals 12 years of age and older. Therefore, the survey procedures are very different for the SAMHSA survey and the Miller et al. (2007) YRBS survey. The SAMHSA surveys are administered through households and do not focus on teenagers whereas the YRBS surveys are administered through schools and entirely focus on teenagers. For the 2004 SAMHSA survey, personal and self-administered interviews were completed for a total of 67, 500 respondents. By the nature of its scope, it provides the best single description of frequency and quantity of drug and alcohol use among a broad age range of people in the United States and the data on adolescent drinking is revealing. In 2004, the data revealed that 3.2% of males and 2.1% of females between the ages of 12 and 17 self-reported heavy alcohol use/binge drinking (five or more drinks per occasion on five or more days in the past 30 days). Note that this definition is different from the YRBS survey where the definition of binge drinking was five or more drinks per occasion on one day, rather than five or more days, within the past 30 days.

The SAMHSA data also demonstrates age trends, with the 18 to 25 age group revealing that 21.2% of males and 8.8% of females were heavy alcohol users. Furthermore, the problem of binge drinking significantly declines with age, with only 9.8% of males and 2.7% of females over 26 years of age reporting heavy alcohol use. When race/ethnicity is taken into account, white survey respondents drink much more heavily than black or Hispanic respondents. For example, among teenagers (ages 12 to 17), 3.3% of whites were heavy alcohol users compared to 2.7% of Hispanics and 0.6% of blacks. This pattern continues with the 18-25 age demographic with 19.4% being heavy alcohol users, followed by 9.0% of

Hispanics and 5.9% of blacks. Finally, in the 26 years or older demographic, 6.7% of whites are heavy alcohol users, while Hispanics and blacks are equivalent at 4.8% each (SAMHSA, 2005).

Binge drinking among high school students is associated with various consequences and numerous health risk behaviors. Miller et al. (2007) reported that students who binge drink do more poorly in school. Furthermore, binge drinkers also engaged in health risk behaviors such as riding with a driver who had been drinking, being currently sexually active, smoking cigarettes, being a victim of dating violence, attempting suicide, and using illicit drugs. A strong dose-response relationship is observed with these associations. The more frequently a high school student reports binge drinking, the higher the prevalence of these other health risk behaviors (Miller et al., 2007).

BINGE DRINKING IN ADOLESCENTS IN OTHER COUNTRIES

As of this writing, there are no comparative studies that directly contrast binge drinking rates of U.S. high school students with their same age peers in other countries. However, there are several studies of binge drinking in adolescents in other countries which seem to indirectly suggest that U.S. teenagers are more likely to binge drink compared with other countries (a trend that has also been shown with college students in the U.S. compared with students in other countries). Feldman et al. (1999) investigated alcohol use among Canadian high school students by administering a questionnaire to 1236 students from 62 randomly selected classrooms in three Canadian urban schools in the year 1994. All of these classrooms were from the Toronto metropolitan region (current population of approximately 5,000,000, which is similar in size to the U.S. city of Atlanta, GA). In this Canadian study, only 11% of the teenagers reported binge drinking (five or more drinks on one occasion at least once a month), a significantly lower number than the 29% of teenagers who reported binge drinking (using the same definition) in the YRBS survey of U.S. high school students. Again, not surprisingly, alcohol use and binge alcohol use rose significantly between grades 9 and 12 and the self-reported binge drinkers were also more likely to engage in high-risk behaviors such as drinking and driving, being a passenger in a car when the driver is intoxicated and daily smoking.

The binge drinking rate of 11% of Canadian high school students is similar to the rate obtained in urban Chinese high school students. Data on alcohol use and other behavioral health risk factors were obtained using the 2004 China Adolescent Behavioral Risk Factor Survey. This survey gathered information about frequency and patterns of alcohol use among adolescents from 18 provincial capitals in China. The impressive data set of 54,040 students comprised of respondents from grades 7 to 12 and the results indicated that 10% of students reported at least one episode of binge drinking in the past 30 days. Prior to the interview, 25% of students reported consuming at least one alcoholic drink in the past 30 days. Binge drinking was more common in males compared to females (14% v. 7%). The Chinese gender difference differs from U.S. estimates of binge drinking where the gender difference is minimal and the gap between males and females has been closing. Despite lower rates of binge drinking in urban Chinese high school students compared to U.S. high school students, 30% of Chinese reported drinking before 13 years of age (Xing et al., 2006). In the U.S., evidence of early drinking (before the start of high school) is often a strong predictor of becoming a heavy binge drinker and/or alcohol dependent by adulthood (Pasch et al., 2008). In the Chinese study, binge drinkers were also more likely to engage in other risky behaviors such as smoking, using drugs, and fighting (Xing et al., 2006), a pattern also observed with adolescents in Brazil (Carlini-Marlatt et al., 2003) and very similar to that seen with U.S. high school students. As such, while binge drinking rates may differ in various countries around the world, the impact of this behavior on health remains the same. Nevertheless, it does appear that while binge drinking in adolescents occurs in other countries, the magnitude of the binge drinking problem is larger in the U.S.

While binge drinking during adolescence may pose significant immediate health hazards to the individual who engages in this behavior, the individual's future health may also be at risk. In order to determine the outcome in adult life of binge drinking in adolescence, longitudinal studies would be the most effective way of measuring this link. Such a study was performed in the UK over the past 3 decades. The 1970 British Birth Cohort Study surveyed participants who were born in 1970 at both age 16 (1986) and at age 30 (2000). With an impressive retention rate (a major concern with longitudinal studies where a significant proportion of the participants are lost), a total of 11,622 subjects participated at age 16 years and 11,261 subjects participated at 30 years. At 16 years, binge drinking was measured in all respondents (defined at two or more episodes of drinking four or more drinks in a row in the previous two weeks). At this time, 18% of participants were binge drinkers. Adolescent binge drinking increased the risk of numerous problems at age 30, including alcohol dependence, excessive regular consumption of alcohol, illicit drug use, psychiatric problems, homelessness, problems with the law, poor academic qualifications, accidents, and lower adult social class. Increased risk for these problems was found even when the authors adjusted for adolescent socioeconomic status (a known risk factor for these various problems). This pattern was not observed to the same degree when examining alcohol consumption patterns that were not binge in nature (i.e., more moderate alcohol use). Thus, adolescent binge drinking (as least the way it was measured in this study) is associated with significant later risk of adversity and social exclusion in adulthood (Viner and Taylor, 2007).

BINGE DRINKING IN COLLEGE STUDENTS

For many college students, attending parties with significant amounts of binge drinking is as important to their college experience as attending their Introduction to Psychology course and learning about Pavlov and his drooling dogs. While consumption of alcohol has long been part of the college experience, the amount of irresponsible and excessive drinking appears to have escalated in the last 20 years. Even in the past, college students always drank more than the general population, but the amount of drinking and the problems that coincide with this behavior have risen dramatically.

Abuse of alcohol on colleges campuses can be chronic (the fraternity member who drinks nearly every day for the four, five or possibly six years that it takes him to acquire his undergraduate degree) or sporadic (e.g., the case of Sam Spady, who drank in excess during a party associated with the big game of the season). When studies have attempted to estimate the rates of binge drinking in college students, the number of 40-50% of all students tends to recur (depending on methodology and how a binge drinker is defined, see Chapter 4). For example, O'Malley and Johnston (2002) estimated that 40% of college students are heavy drinkers by their definition of having five or more drinks in a row in the past two weeks.

Harold Wechsler and his colleagues at the Harvard School of Public Health have provided the most comprehensive data on rates of binge drinking on college campuses in the United States through their national College Alcohol Studies (CAS). Large surveys of more than 50,000 students at 140 four-year colleges in 40 states were completed in 1993, 1997, 1999, and 2001. Schools were selected to be representative of all colleges and universities in the United States by including schools that were public and private, urban and rural, and of all sizes and academic competitiveness. A nineteen page questionnaire was sent to a random sample of more than 200 undergraduates at each participating institution. The students answered yes/no and multiple choice questions about their drinking and related problems associated with drinking. Wechsler defined the term "binge drinking" as drinking five or more drinks in a row at least once in the prior two weeks for men and drinking four or more drinks in a row for women. This cut-off was based on the examination of problems associated with different levels of alcohol intake following the first study in 1993. Note that Wechsler's definition did incorporate gender as part of the definition of binge drinking to account for the overall smaller body size and metabolism rate of alcohol in women. Wechsler further classified students as "occasional binge drinkers" if they reported binge drinking in this manner once or twice in the previous two weeks, and "frequent binge drinkers" if they reported binge drinking three or more time in the previous two weeks (Wechsler et al., 2002).

In the most recent 2001 survey, 81% of women and 79% of men reported that they drank alcohol in the last year. Furthermore, 41% of women and 49% of men met the definition of "binge drinkers". The gender difference of slightly higher rates of binge drinking in males, while not large, was statistically significant. There were also important trends revealed when rates of binge drinking from the most recent survey are compared with the first survey in 1993. The percentage of binge drinkers has remained essentially the same over time. However, a polarization in drinking behavior appeared to have occurred. The percentage of abstainers (i.e., individuals who reported that they did not drink any alcohol in the last year) increased in 2001 compared with 1993, but the individuals who binge drank often also was increasing. It is important to note that with the CAS surveys, the average results do not apply to all students or all campuses. Students from religious schools, commuter schools, and historically Black colleges and universities drank much less. By contrast, first-year students, Whites, members of fraternities and sororities and athletes drank much more (NIAAA, 2002).

What are the consequences of college binge drinking? The Task Force of the National Advisory Council on Alcohol Abuse and Alcoholism (2002) attempted to provide an indication of the sometimes staggering consequences of misuse of alcohol among college students in the United States. This task force reviewed all of the available studies on college student drinking to make recommendations about how to change the culture of drinking on college campuses. In their report, *A Call to Action: Changing the Culture of Drinking at U.S. Colleges,* the reported consequences of excessive drinking ranged from benign to fatal. Of the estimated 6 million college students in the United States, about 25% of the students reported academic consequences of their drinking. These problems included missing classes, falling behind, doing poorly on an exam or paper, and receiving lower grades overall (Engs et al., 1996; Presley et al., 1996a,b; Wechsler et al., 2002). This behavior can also be problematic for school administrators. When school administrators were asked about property damage, more than 25% of administrators from schools with relatively low drinking levels and over 50% from schools with high drinking levels say their campuses have significant problem with alcohol-related property damage (Wechsler et al., 1995). The reports of school administrators coincided with questionnaire based self-reports by students. About 11% of college student drinkers reported that they had damaged property while under the influence of alcohol (Wechsler et al., 2002).

Drinking to get drunk also results in serious health consequences. Each year it is estimated that more than 150,000 students develop an alcohol-related health problem and 99,000 students are unintentionally injured while under the influence (Hingson et al., 2005). In addition, the relationship between binge drinking and unsafe sexual behavior has been well-documented. It has been estimated that 400,000 students had unprotected sex while drinking, and more than 100,000 students reported having been too intoxicated to know whether they had consented to having sex (Hingson et al., 2002). Surveys also indicate that between 1.2% and 1.5% of all students said that they tried to commit suicide within the past year due to drinking or drug use (Presley et al., 1998). There is also evidence to support that binge drinking leads to more serious alcohol abuse and dependence problems. Some binge drinkers mature out of this behavior upon leaving college, though a subset of individuals acquire an alcohol dependence problem during college. Based on data from questionnaire-based self-reports about their drinking, it is estimated that 31% of college students met criteria for a diagnosis of alcohol abuse, and 6% for a diagnosis of alcohol dependence in the past 12 months (Knight et al., 2002).

Binge drinking on college campuses results in significant legal troubles for students who engage in this behavior. For example, estimates indicate that almost 700,000 students per year are assaulted by another student who has been drinking. Further, more than 97,000 students reported being victims of alcohol-related sexual assault or date rape in a recent year (Hingson et al., 2005). Police reports indicate that obeying the law diminishes as drinking increases. About 1 in 20 four-year college students who drink get involved with the police or campus security as a result of their drinking (Wechsler et al., 2002). In a detriment to future job prospects where arrests must be disclosed, an estimated 110,000 students between the ages of 18 and 24 are arrested for an alcohol-related violation such as public drunkenness or driving under the influence each year (Hingson et al., 2002).

However, it is the rate of impaired driving among college student drinkers that is most staggering. A 2002 survey indicated that 2.1 million college students between the ages of 18 and 24 drove under the influence of alcohol in the prior year (Hingson et al., 2002). Considering that there are approximately 6 million college students in the U.S., this means that about 1/3 of college students drive drunk at least once per year. That statistic alone is frightening and should dissuade anyone who thinks that the problem of binge drinking in college students is innocuous and nonthreatening to the well-being of our society as a whole.

Perhaps most tragic are the deaths that result from high-risk college drinking. Approximately 1,700 college students between the ages of 18 and 24 die each year from alcohol-related unintentional injuries, including motor vehicle crashes (Hingson et al., 2005). Deaths and serious injuries due to binge drinking are the main reason why heavy drinking among college students has drawn significant societal attention. Although binge drinking during college increases the likelihood for future problems, we cannot predict which students will have alcohol problems after they leave college. While research has shown that significant percentages of people who do have alcohol problems as adults drank heavily in college, most people who drink heavily in college do not have later alcohol problems (Maisto et al., 2008). Thus, the greatest problem with binge drinking in college students may be the risk to health and well-being for the students while they are students.

BINGE DRINKING IN COLLEGE STUDENTS IN OTHER COUNTRIES

While binge drinking has been under the greatest investigation in the United States, several studies have been conducted regarding rates of binge drinking among college students in other countries. Canada has a college student population that is similar to that in the United States, although the age that an individual can legally drink is lower in most Canadian provinces. The most recent data on the prevalence and frequency of binge drinking (heavy drinking episodes) among Canadian undergraduate students was gathered with the Canadian Campus Survey in 1998. This survey was a national mail survey sent to a random sample of 7,800 students from 16 universities. Students were asked about consuming 5 or more drinks or 8 or more drinks per occasion (there was no threshold change for gender). Overall, 63% of students reported 5 or more drinks at least once in the past semester and 35% of students reported 8 or more drinks at least once in the past semester. Those students who reported 5 or more drinks at least also reported that they repeated that consumption pattern almost 5 times during the semester. Students who reported 8 or more drinks reported that they repeated that pattern twice during the semester. The gender difference seen with U.S. college students was also found, with males having higher rates of heavy drinking. Heavy drinkers also were more likely to live in university residences and exhibit low academic performance (Gliksman et al., 2003).

To compare the patterns of drinking between the United States and Canada, the data from the 1998 Canadian Campus Survey was compared with the data obtained from the 1999 College Alcohol Survey.

Prevalence of life-time and past year alcohol use was found to be higher among Canadian students than U.S. students (92% versus 86% for life-time alcohol use, 87% versus 81% for past year use). However, the prevalence of binge alcohol use (5 or more drinks per occasion for males and 4 or more drinks per occasion for females) for past-week and past-year drinkers was significantly higher among U.S. students compared to their Canadian counterparts (41% versus 35% for past-week binge alcohol use, 54% versus 42% for past-year binge alcohol use). Hence, the authors found that while Canadian students were more likely to drink, it was the American students who drank more and in excess. The differences are not dramatic, as the numbers indicate that there was still a significant pattern of binge drinking in the Canadian students. As for demographic considerations of who drinks more or less, the authors found that in both countries, older students drank more moderately as did students who lived at home with their parents. Furthermore, students who reported getting drunk for the first time before the age of 16 were more likely to be binge drinkers in college (Kuo et al., 2002).

SUMMARY

The most recent epidemiological studies of alcohol consumption patterns of adolescents and college students have found that binge drinking, or drinking to get drunk, is common. While studies differ in their estimates of binge drinking depending on how this behavior is defined, in general it appears that almost one third of high school students and almost one half of college students appear to regularly binge drink. The consequences of this behavior to health and well-being are significant. Those individuals who binge drink tend to have poorer academic performance and engage in numerous other health risk behaviors. Driving while impaired, engaging in unsafe sexual behavior and using illicit drugs are some of the health-risk behaviors associated with binge drinking. Furthermore, there are many preventable accidental deaths associated with binge drinking. Binge drinking among college students is clearly a major problem to be addressed, with long-lasting and potentially permanent effects.

Chapter 4

DEFINITION OF BINGE

One of the difficulties that professionals who study binge drinking are grappling with is: What is a binge and who should be called a binge drinker? This seems to be such a simple problem to solve. It would appear that a consensus should be reached, so progress can be made in understanding whether binge drinkers experience greater deleterious health and social effects due to their drinking and how their consumption patterns can be moderated to a safer range. Unfortunately, the field has so far been unable to decide on: 1) what is a binge and 2) who we should call a binge drinker. The material on epidemiology presented in Chapter 3 gave a preview of this problem. To determine rates of binge drinking in adolescents and college students, one must have an operational definition of a binge. However, there are almost as many definitions of 'binge' and 'binge drinker' as there are studies of binge drinking. This creates problems in the literature. Scientists and physicians need to know the scope of the binge drinking problem in various demographic groups such as high school students and college students. They also need to be able to identify improvements over time if new prevention programs are implemented. Cross-cultural comparisons of binge drinking rates are similarly difficult to execute since there is not only little consensus on the definition of a binge drinker in the United States, but researchers in other countries also appear to be similarly paralyzed by this problem.

DEFINITION OF A STANDARD DRINK

Before the available definitions of binge drinking can be compared, it is helpful to define a standard drink, since even a standard drink is not the same in every country. In the United States, a standard drink contains 13.7 grams (0.6 ounces) of pure alcohol. Thus, a 12 ounce beer, 8 ounces of malt liquor, 5 ounces of wine, or 1.5 ounces or a "shot" of 80-proof (40% alcohol per volume) distilled spirits or liquor (such as gin, rum, vodka, whiskey etc.) would all be considered standard drinks (Department of Health and Human Services Centers for Disease Control and Prevention, http://www.cdc.gov/alcohol/faqs.htm). The current definition of a standard drink is not without some concerns. For example, 12 ounce beers can differ substantially in alcohol concentration. Logan et al. (1999) measured the alcohol content of 404 beers and malt beverages available for sale in the state of Washington, three-quarters of which were brewed in the United States. Alcohol concentrations ranged widely from 2.92 to 15.66% per volume and, surprisingly, there were often disparities between the stated alcohol content on the label of the bottle and the true alcohol content of the beverage. Thus, individuals may be consuming significantly different amounts of alcohol depending on the brand of the beverage and the reliability of the manufacturer to provide a product with the stated alcohol content. Furthermore, the average alcohol concentration of beers has risen over time. From 1995 to 2000, the average alcohol content of a beer in the U.S. rose from 4.33 to 4.66 % per volume (Kerr and Greenfield, 2003). The total volume of beer consumed in the U.S. between 1995 and 2000 dropped which led some to suggest that alcohol consumption among beer drinkers decreased during this period. However, if the change in alcohol content is also considered, analyses revealed that overall alcohol consumption from beer actually increased during those years (Kerr and Greenfield, 2003).

When comparing studies from different countries, it is important to note that the definition of a standard drink differs across countries. In the United Kingdom, where binge drinking is a serious societal problem and thus a significant amount of binge drinking research is being performed, a standard drink is smaller than in the United States. In the U.K., a standard unit (their commonly used term for a standard drink) contains 8 grams of alcohol (compared with 13.7 grams of alcohol in the U.S.). Using a shot as a comparison, a 25 ml shot is a standard unit whereas a shot in the U.S. is 45 ml (1.5 ounces). Thus, in England, researchers often define binge drinking as 8+ units for men and 6+ for women (Department of Health, 2007). These units of alcohol coincide with approximately 4.7+ standard drinks for men and 3.5+ standard drinks for women in the United States.

DEFINITION OF BINGE

The variability of available definitions for binge drinking is partially due to the slow transition from considering binge drinking as a "rite of passage" to a health risk behavior. In the past, researchers were concerned with accurately defining alcohol abuse and dependence (as described in Chapter 2). Thus, standardized screening instruments for problem drinking are often weighted toward physical symptoms of alcohol dependence and thus have limited utility in college and high school populations. For example, the CAGE (Cut Down, Annoyed, Guilty, Eye-opener) questionnaire is based on the disease model of alcoholism and adult conceptualizations of alcohol-related problems. The four interview questions include: 1) Have you ever felt you ought to cut down on your drinking, 2) Have people annoyed you by criticizing your drinking, 3) Have you ever felt bad or guilty about your drinking, and 4) Have you ever had a drink first thing in the morning to steady your nerves or get rid of a hangover (eye-opener). Even heavy binge drinking college students or high school students will answer most of these questions in the negative. Moreover, the culture of college drinking does not typically encourage any feelings of guilt about overindulgence, and instead encourages the opposite, making this questionnaire inappropriate for a college-age demographic group (O'Brien et al., 2006).

Thus, the realization that binge drinking is neither innocuous nor alcohol dependence has led researchers to struggle with how to define binge drinking. Specifically, the challenge presented to researchers is to develop a definition that does not include those individuals who don't have a problem with their alcohol consumption patterns, while also not including those individuals who are seriously abusing alcohol and are in need of treatment.

For example, in England, the Prime Minister's strategy unit defined binge drinkers as individuals who drink "to get drunk" (Prime Minister's Strategy, 2004). On the surface, classifying individuals who drink to get drunk seems to be a sensible and simple approach to categorize drinkers into binge and moderate groups. For example, O'Brien et al. (2006) asked a sample of 3,909 U.S. college students the simple question, "In a typical week, how many days do you get drunk?" The authors found that 54% of students reported getting drunk at least once in a typical week. Furthermore, students who got drunk at least once per week were also more likely to be hurt or injured as a result of their own drinking, experience a fall from a height that required medical treatment, or be taken advantage of sexually as a result of another's drinking, compared to students who didn't get drunk. Similarly, students who drank to get drunk also were also more likely to *cause* an injury requiring medical treatment to someone else, were more likely to cause an injury in an automobile crash, cause a burn that required medical treatment, or cause a fall from a height that required medical treatment.

However, defining binge drinking as getting drunk has problems. How often do you have to get drunk to make you a binge drinker? According the UK definition, frequency doesn't matter. If you got drunk once a year, are you a binge drinker? What about once a week (as in the O'Brien et al., 2006 study described above)? The second problem concerns the term 'drunk'. What exactly is getting drunk? Is it the feeling of being slightly light-headed after one or two drinks? Is it passing out after 12 drinks? Each individual seems to approach the term "drunk" with preconceived notions based on their own experiences and watching the experiences of those around them. Concepts about what "drunk" looks and feels like differ dramatically among individuals.

As such, many researchers have tried to operationalize the definition of binge by including standard drinks in the definition. Using standard drinks, individual response differences to the question posed can be somewhat reduced. However, even this approach has some problems.

When asked a question such as "do you typically drink 5 or more drinks on one occasion?" several individuals may answer yes. However, these individuals may differ on multiple important characteristics which determine whether they are getting intoxicated. As an example, body weight, gender and duration of time spent drinking that amount of alcohol are factors that matter a great deal. A 120 pound female will probably be intoxicated by 5 standard drinks on a particular occasion. However, a 230 pound male may not. Even the small woman may not be intoxicated if her duration of drinking was over a full evening of 6 hours and drinking was spaced out evenly over that time period.

Finally, the wording of the question needs to be considered carefully, which is why some definitions of binge have been more readily embraced than others. In the above question, individuals may wonder what the researcher means as "typical". Does 'typical' refer to how I drink every week or does it refer to only the times when I drink (which might be often or only a few times per year)? For these reasons, consensus criteria of what a binge should be defined as have yet to be embraced. Some of the variance in definitions of binge and binge drinker being used by various researchers can be seen in Table 4-1.

Table 4-1. Definitions of Binge Drinking. Rates of binge drinking are stated if reported in a study

Definition	Population	Corrected for gender?	Rates	Studies
Prolonged period of intoxication or excessive heavy drinking that can last for days or weeks (also called a bender); in reference to a person in the chronic phase of alcoholism	Alcoholics	No	Not applicable	Jellinek (1952)
5 or more drinks per occasion for men during the past 2 weeks; 4 or more drinks per occasion for women during the past 2 weeks	U.S. college students (Harvard School of Public Health College Alcohol Studies conducted in the years 1993, 1997, 1999, 2001); Centers for Disease Control and Prevention Behavioral Risk Factor Surveillance System	Yes	44%-53% for college students (~50% for men, ~40% for women); 15% nationwide for all U.S. respondents aged 12 and older in 2007	Wechsler et al. (1994, 1995, 1997, 1998, 2000, 2002, 2006); Wechsler and Isaac (1992); Wechsler and Nelson (2001); Cranford et al. (2006); see www.cdc.gov for state by state U.S. annual prevalence data statistics on binge drinking
5 or more drinks during the past 2 weeks	U.S. college students (University of Michigan's Monitoring the Future Project)	No	41%	Johnston et al. (1991)
5 or more drinks during the past 2 weeks	U.S. college students (Core Alcohol and Drug Survey)	No	42%	Presley et al. (1993)
Drinking resulting in a peak blood alcohol concentration of .08g% or above (5 or more drinks during a 2 hour period for men; 4 or more drinks during a 2 hour period for women)	Definition endorsed by the National Institute on Alcohol Abuse and Alcoholism (NIAAA, 2004)	Yes	64% (assuming at least one binge in the past year – Cranford et al., 2006)	Cranford et al. (2006); D'Onofrio et al. (2008); Marczinski et al. (2007)

Definition	Population	Corrected for gender?	Rates	Studies
5 or more drinks per occasion during the past 30 days	U.S. high school students (2003 National Youth Risk Behavior Survey)	No	29%	Miller et al., (2007)
5 or more drinks per occasion during the past year	U.S. adolescents ages 11 to 18 years (Growing Up Today Study)	No	29% of male teens and 24% of female teens who initiated alcohol use	Fisher et al. (2007)
5 or more drinks on a single occasion during one semester of school	Canadian college students (Canadian Campus Survey)	No	63%	Gliksman et al. (2003)
8 or more drinks on a single occasion during one semester of school	Canadian college students (Canadian Campus Survey)	No	35%	Gliksman et al. (2003)
8 or more U.K. standard drinks per occasion for men; 6 or more drinks per occasion for women	Citizens of the United Kingdom	Yes	32% of men; 9% of women	U.K. Department of Health (2008); Drummond et al. (2005)
Drinking to get drunk	Citizens of the United Kingdom	No	Not estimated	U.K. Prime Minister's Strategy Unit (2004)
Drunk at least once in a typical week	U.S. college students	No	54%	O'Brien et al. (2006)
4 or more drinks per occasion at least twice in the previous 2 weeks	1970 British Cohort Study participants (all age 16)	No	18%	Viner and Taylor (2007)
Calculated from the Alcohol Use Questionnaire (4xSpeed of Drinking + number of times being drunk in the previous 6 months + 0.2xPercentage of times getting drunk when drinking), score of 24+ for binge drinkers	UK college students	No	Not estimated	Mehrabian and Russell (1978); Rose and Grunsell (2008); Townshend and Duka (2005)

The variability in use of the terms 'binge' and 'binge drinker', as shown in Table 4.1, is dramatic. The variability in the use of the term 'binge' results in drastically different rates of binge drinking for high school and college students (18-64%). All of the researchers above would state that their use of the term binge refers to a threshold at which drinking becomes associated with other health harms and risks. For example, the widely used Wechsler criteria of 5 or more drinks per episode for men (4 or more drinks per episode for women) in the past 2 weeks is an appropriate threshold as those individuals who meet or exceed that number of drinks are more likely to get into automobile accidents when drinking, have unprotected sex when drinking, get into fights, display physical or cognitive impairment, and have poor academic performance. Thus, this threshold provides a line at which consumption of a sufficiently large amount of alcohol places the drinker at increased risk of experiencing alcohol-related problems and/or places others around him or her at increased risk of experiencing secondhand effects (Wechsler and Nelson, 2001).

One notable aspect of the variability in the definitions of binge drinking is that some researchers have corrected for gender while others have not. It has been shown that women experience alcohol-related problems at lower drink levels than men. This has been the case even when body mass was controlled, due at least in part because of gender difference in enzymes involved in the metabolism of alcohol (Wechsler et al., 1995). However, we don't yet know whether gender differences in alcohol metabolism rates and body fat percentages impact teenage and college students differently.

It is worth noting that part of the problem in embracing one universal definition of binge and binge drinker is that there are still many substance abuse researchers who think that the term 'binge' is inappropriate for high school or college students. Clinicians have historically reserved the term binge to refer to the drinking pattern of a person in the chronic phase of alcoholism. In this case, the drinking binge (also called a bender) is a prolonged period of intoxication or excessive heavy drinking that can last for days or weeks (Jellinek, 1952). As such, the clinicians and researchers object to the use of the term binge in reference to high school and college student drinking behavior, because among this younger group, a comparatively smaller number of drinks consumed in a shorter time span is involved. Similarly, it is the policy of some journals in the field, such as the *Journal of Studies on Alcohol and Drugs*, to limit the use of the terms "binge" and "binge drinking" to an extended period of time (usually two or more days) during which a person repeatedly administers alcohol or another substance to the point of intoxication, and gives up his/her usual activities and obligations in order to use the substance. It is the combination of prolonged use and the giving up of usual activities that forms the core of this definition of a binge. Therefore, authors wanting to describe heavy college drinking must use the terms "heavy drinking" or "heavy episodic drinking" in order to publish in that journal (see http://www.jsad.com/jsad/static/binge.html).

However, the term binge is slowly being embraced by many researchers and some clinicians as the appropriate term to describe the heavy episodic consumption of alcohol typical of a college student. The exact reason for this change in approach is somewhat unclear. However, part of the change may be due to the fact that the word binge is now not just the domain of substance abuse researchers and clinicians. Binge is used by the general public for any period of unrestrained, immoderate, excessive or uncontrolled self-indulgence and the time spent on the activity often doesn't matter. An eating binge could refer to heavy caloric intake over a short period of time. A shopping binge may refer to an afternoon of unrestrained spending. Thus, the increased use of the term binge to describe the type of heavy episodic drinking of college students (Wechsler and Austin, 1998; Wechsler and Nelson, 2001) may help the general public better understand the nature and scope of this problem. At the very least, the media has embraced this definition, as its usage has increased dramatically and the term binge has become the primary way this form of drinking by college students is identified. For example, Wechsler and Nelson (2001) reported almost no media mentions of binge drinking starting in the early 1990s but close to 150 media mentions annually by the late 1990s, using the Lexis-Nexis Academic Universe database of 32 major newspapers in the United States.

WHAT YOUNG PEOPLE CONSIDER A BINGE

Researchers and clinicians continue to squabble over the appropriate definition of binge drinking. However, it would be interesting to know what high school students and college students actually consider to be binge drinking. Wechsler and Kuo (2000) asked college students to define the term binge drinking, as part of their larger College Alcohol Study. The authors found that the students' definition of binge drinking was 6 drinks in a row for men and 5 in a row for women. This threshold is 1 drink higher than the 5/4 definition usually used by many researchers in the field. Not surprisingly, the definitions generated by the students varied with their own drinking levels (i.e., heavier drinkers generated higher thresholds of binge drinking). The students were also asked to estimate how many students at their school were binge drinkers. The median estimate was that 35% of students were binge drinkers (compared to the median value of 44% generated from the College Alcohol Studies). Half of the students underestimated the binge drinking rate at their school, 30% overestimated it and only 13% were accurate.

ACCURACY IN REPORTED ALCOHOL CONSUMPTION

For all of the above studies cited in Table 4.1, rates on binge drinking were compiled from survey data. Adolescents or college students were asked to think back to their recent drinking behavior and estimate how many standard drinks they had consumed during those drinking episodes. The task of accurately assessing past consumption levels is not as reliable as one might hope. Problems that have been raised with these alcohol surveys in college students include: survey respondents lacking the necessary knowledge of standard drink volumes to properly answer the questions in the survey, and misperceptions of standard serving sizes leading to inaccuracies in self-reported consumption. For example, White et al. (2005) asked students to complete a questionnaire that queried standard drink volumes, and they also asked students to complete a task where they free-poured a single beer, glass of wine, shot of liquor, or amount of liquor to make a mixed drink. With the exception of the beer, students incorrectly defined the volumes of standard servings of alcohol by overestimating the appropriate volumes. When free pouring the drinks, they also overestimated the appropriate volumes. This overestimation became worse when students were pouring drinks into larger containers. Thus, it is possible that college students are often under-reporting their alcohol consumption on surveys (White et al., 2003, 2005). Inaccuracies in knowledge of standard drink volumes may not be so surprising considering that the experience with alcohol for many college students is largely with free poured drinks and beer from kegs. However, it appears that poor knowledge of standard drink volumes may not be limited to younger and less experienced drinkers. Gill and Donaghy (2004) studied individuals from the general Scottish population, asking their subjects to define standard drink volumes and then free pour a single serving of wine or spirits. The authors found that their subjects poured a drink of wine that corresponded to 1.9 standard servings, and a drink of liquor that corresponded to 2.3 standard servings. Thus, it appears that poor knowledge of standard drink volumes is a pervasive problem across age groups.

As such, studies that use thresholds for binge drinking to sort students into categories on the basis of their drinking habits might be vulnerable to the poor knowledge of standard drink volumes. Furthermore, small changes in drinking levels could result in significant changes in the number of students who meet the criteria for binge drinking (Dawson, 1998). As thresholds for binge drinking are based on self-report data, thresholds to indicate risky drinking may themselves be set at inappropriate levels (White et al. 2005).

However, another study suggested that inaccuracies in serving size estimation may not be skewing self-report data as much as originally thought. Underestimation of the amount of alcohol consumed is a problem when students overestimate the appropriate sizes of standard drinks. Thus, Kraus et al. (2005) asked college students to estimate their blood alcohol concentration (BAC) following drinking in a field study. The authors observed that estimated BAC levels were actually significantly higher, not lower, than breath BAC measures. In the midst of drinking, students might actually overestimate rather than underestimate levels of consumption. In a similar type of field study, Beirness et al. (2004) gave 856 Canadian college students who were returning home from a night out between 10 p.m. and 3 a.m., a survey and a breathalyzer test. The survey was used to determine whether the individual was a binge or non-binge drinker using Wechsler's 5+/4+ criteria. In the entire sample, 74% of students had a BAC of zero and 12% had a BAC less than .05%. The authors found that among the students classified as binge drinkers, 49% had a zero BAC on the night they were interviewed. Even those students who had consumed 5+/4+ drinks the evening of the interview had a mean BAC less than .08%. It is unknown if the students who reported that they had binged earlier (i.e., consumed 5+/4+ drinks that night), the earlier binge had resulted in a peak BAC of .08%. Clearly, questionnaire results and behavior may not be matching up in many cases. The results of these field studies fuel an already raging debate about which definition should be applied, and whether the term binge is appropriate in the first place when describing heavy alcohol consumption in young people.

CONCLUSION

There are numerous ways to define binge drinking, and no universal definition of the term 'binge' or 'binge drinker' has become widely accepted as of yet. Definitions of binge drinking vary from the simple, such as drinking to get drunk, to complex calculations that incorporate aspects of quantity and frequency of alcohol consumption. Some definitions correct for gender, whereas others do not. One widely cited definition of a binge drinker is a male who consumes five or more drinks on one occasion within the past two weeks or a female who consumes four or more drinks on one occasion within the past two weeks. Lack of a consensus definition for the terms 'binge' and 'binge drinker' limits cross-study comparisons and slows our understanding of the nature and scope of the problem.

Chapter 5

SOCIAL AND ENVIRONMENTAL FACTORS AND BINGE DRINKING

Binge drinking in adolescents and college students is a social phenomenon. Environmental factors play an important role, perhaps more than individual characteristics, in the development of drinking and binge drinking in adolescents and college students (Presley et al., 2002). Particularly on many college campuses, a culture of alcohol abuse permeates many campus student organizations despite the efforts of college administrators to discourage this behavior. In this chapter, we examine a variety of social factors that are at play in the binge drinking problem among young people. Drinking depends on time, place, situation and personal characteristics and drinking contexts can vary considerably in terms of social rules and norms for appropriate drinking behavior (Greenfield and Room, 1997). Unfortunately, some contexts (college itself, fraternities/sororities, and college athletics) have drinking norms emphasizing drunkenness that are harmful and encourage and escalate binge drinking.

FAMILY AND PEER NORMS

The drinking patterns of peers are important predictors of drinking patterns in high school and college students. In high school students, friends' drinking is strongly related to personal alcohol use and peer influence is stronger for girls than boys (Dick et al., 2007). Interestingly, peer influences may even be stronger than parental influences. In a study of adolescents from grades 7 to 12, parental monitoring of drinking, attachment to mother and attachment to father have been shown to have significant, albeit small, influences on binge drinking rates in high school students. However, peer influences had strong effects on adolescent binge drinking rates (Bahr et al., 2005). Other studies have also shown that parents with a permissive parenting style (i.e., few rules and expectations) often have children who are more likely to drink heavily in high school (Tucker et al., 2008). It remains unclear why peers matter more and parents matter less. It could be that parents don't make the effort to prevent or intervene in their adolescent's underage drinking behavior. It is possible that parents do make significant efforts to curtail their child's drinking, but their efforts fail because they use an inappropriate approach for their child, or because their child would ignore the message regardless of how it was presented. Not surprisingly, the role of peer drinking in predicting alcohol consumption patterns increases in importance once an individual attends college, when the individual may be away from parental influences or where parents no longer exert strong control over their child's environment (Jamison and Myers, 2008).

COLLEGE

Attending college is a social factor that increases the likelihood of binge drinking. Some college binge drinkers are just continuing a behavioral pattern that was already established in high school. Other college students did not binge drink in high school and thus can begin or avoid binge drinking in college. Weitzman et al. (2003) investigated the factors associated with initiation of binge drinking in a national sample of college students. They observed that the college students who were exposed to more drinking environments (e.g., social, residential, and market surroundings in which drinking is prevalent and alcohol is inexpensive and easily accessed) were more likely to acquire the pattern of binge drinking. Findings from this study would suggest that reductions in college binge drinking could result from efforts to limit access to or availability of alcohol, controlling prices of alcohol, and maximizing substance free environments and associations.

Presley et al. (2002) reviewed the literature on college drinking to examine the aspects of the college environment (e.g., presence of fraternities/sororities), rather than the student characteristics (e.g., genetic predisposition to drinking problems) that influence drinking. They noted several environmental factors associated with drinking. *Organizational property variables* on campuses are associated with drinking, such as affiliations (historically black institutions and women's institutions have less binge drinking), presence of a Greek system (presence of fraternities and sororities is associated with more binge drinking), and athletics (college athletes are more likely to binge drink). Second, the *physical and behavioral property variables* of college campuses influence college drinking. Included are types of residence, institution size, location and quantity of heavy binge drinking. Finally, *campus community property variables* are associated with college drinking. These include the price of alcohol, availability of alcohol, and outlet density of places that sell alcohol. While the campus environment is a complex place, there are controllable factors that can increase or decrease binge drinking.

COLLEGE FRATERNITIES AND SORORITIES

Fraternity and sorority members use more alcohol than nonmember peer college students. Fraternity and sorority members not only consume more drinks per week, but they also are more likely to binge drink and suffer negative consequences from their drinking (e.g., suffer injuries) compared to their nonmember peers. The leadership of these organizations appears to participate in heavy drinking and set and encourage heavy-drinking norms. Individuals with leadership positions consume more alcohol, engage in binge drinking and experience negative consequences at levels at least as high, and in some cases much higher, than that of their members. When fraternity or sorority members were queried about their views about alcohol, members reported that alcohol was a vehicle for friendship, social activity and sexuality, to a greater extent than for nonmember college peers (Cashin et al., 1998).

Environmental and social context may be driving the binge drinking patterns of fraternity and sorority members. In an interesting longitudinal study of college drinkers, Bartholow et al. (2003) observed that while affiliation with a fraternity or sorority is an important risk factor for heavy drinking, this risk may be limited to the college years. In their sample, they observed that greater cumulative exposure to fraternities or sororities led to increased heavy drinking during the college years. Shortly after leaving college, heavy drinking levels plummeted and remained low through approximately age 30 (Bartholow et al., 2003). Thus, peer norms for heavy drinking may mediate the observed relationship between fraternity/sorority affiliation and heavy drinking, emphasizing the likelihood that situational determinants play an important role in binge drinking in young people (Sher et al., 2001). However, other studies have demonstrated the importance of self-selection in fraternity/sorority membership. Members are often heavier substance users before starting college and may be attracted to organizations that are associated with heavy drinking (Park et al., 2008). As such, fraternity and sorority members may constitute an at-risk group for problem drinking prior to entering college (Capone et al., 2007).

COLLEGE ATHLETICS

College athletes report more binge drinking, heavier alcohol use, and a greater number of drinking-related harms, compared to their non-athlete peers (Brenner and Swanik, 2007; Martin, 1998; Nelson and Wechsler, 2001; O'Brien et al., 2008). Not surprisingly, college athletes are exposed to more alcohol prevention efforts since their drinking may hamper their athletic performance. Despite exposure to prevention programs, athletes are still a high-risk group for binge drinking and alcohol-related harms (Nelson and Wechsler, 2001). This risk also extends within the college athletics community to sports fans, who have also been found to be more likely to binge drink (Nelson and Wechsler, 2003). When different sports are examined, athletes involved in individual sports are less likely to drink heavily in comparison to athletes that are involved in team sports (Brenner and Swanik, 2007). Athletes alter their drinking behavior dependent on the demands of their sports. Martin (1998) investigated alcohol use among National College Athletic Association Division 1 female college basketball, softball, and volleyball athletes. The athletes moderated their drinking during the competitive sports season, with binge drinking rates of 35% in season increasing to 60% out of season. When athletes were asked to give reasons why they did not use alcohol, they reported reasons such as the effects alcohol has on health and sports performance, coaches' rules, and concerns about weight gain. When asked for reasons for using alcohol, most athletes stated that they drank for social reasons.

MINIMUM LEGAL DRINKING AGE 21 (MLDA-21)

In 1984, the National Minimum Drinking Age was passed by the United States Congress as the mechanism whereby all states would be required to legislate and enforce the minimum age for purchasing and publicly possessing alcoholic beverages at 21 years. To ensure that the 21 year old minimum drinking age would be enforced, each state would be subjected to a

10% decrease in its annual federal highway apportionment under the Federal Aid Highway Act for non-compliance. Most states still permit "underage" consumption in limited and specific circumstances (e.g., during religious occasions) (NIAAA, retrieved from the Alcohol Policy Information System September 30, 2008).

Since drinking is now illegal for all individuals under the age of 21 in all U.S. states, zero tolerance laws for driving have also been instituted. Thus, a driver who is under the age of 21 who registers a blood alcohol level of .02 g% (or .01 g% depending on the state law) will suffer harsh penalties for their behavior. Some research has determined the extent to which observed declines in alcohol-related highway deaths among drivers younger than age 21 years can be attributed to the change to MLDA-21 and establishing zero tolerance laws. Voas et al. (2003) compiled data on all drivers younger than age 21 years involved in fatalities in the U.S. between 1982 and 1997. They observed substantial reductions in alcohol-positive involvement in fatal crashes associated with these two laws that are specific to youth, suggesting that the policy of limiting alcohol access to youth through MLDA laws and reinforcing this action by making it illegal for any underage driver to have any alcohol appears to be effective in saving lives (Fell et al., 2008; Voas et al., 2003).

Raising the drinking age to 21 may have other benefits, including reducing the risk for future problems with alcohol. For example, detailed data on military service persons can be used to evaluate various alcohol control policies. Before 1982, soldiers consumed alcohol legally on U.S. bases, regardless of age. By 1988, the military decided to change their approach to alcohol, similar to changes that were occurring in the civilian population with the enactment of MLDA-21 in 1984. With this change in approach, the military established policies to discourage underage and problem drinking and fully transitioned to a MLDA-21. Change in military policy to alcohol was evaluated by changes in later alcohol treatment episodes among male veterans and civilians between the years 1992 and 2003, with four different age cohorts being examined. Alcohol treatment rates in both the civilian and veteran groups were similar. By 2003, veterans' treatment rates for alcohol problems fell 60% for ages 25 to 34 compared with a 20 to 25% reductions for civilians of the same age. As such, the military's efforts to strongly enforce MLDA-21 appear to result in greater reductions in later alcohol treatment episodes among veterans compared with civilians (Wallace et al., 2008). Similar results have also been found with the general public. States differ in the stringency of their alcohol control policies. Nelson et al. (2005) reported that for students attending college with stringent alcohol control policies (e.g., strict enforcement of MLDA-21, limiting liquor outlet density or mandating keg registration), binge drinking rates were significantly lower than for students attending college in states with more lenient alcohol control policies.

CROSS-BORDER BINGE-DRINKING

While the findings from some research clearly support MLDA-21 and demonstrate that raising the drinking age may offer protection against particular problems such as risk of death from impaired driving or future serious alcohol dependence problems, there are still significant numbers of binge drinking high school and college students who suffer serious consequences because of their drinking (Wechsler et al., 1996, 1998, 2000). The enactment of

MLDA-21 may have pushed problem binge drinking into more hidden environments as youths find ways to escape limits on drinking alcohol. One example is the problem of cross-border drinking (Lange et al., 1999, 2002; Lange and Voas, 2000; Voas et al., 2002, 2006). The MLDA-21 in the United States and a younger drinking age in the neighboring countries of Canada (18/19) and Mexico (18) have led many adolescents who live close to the border to cross country borders in order to drink. This cultural phenomenon has been studied a great deal in southern California where many young people travel to Tijuana, Mexico, which has a loosely-enforced age-18 law, and where the cost of alcoholic drinks is much lower than similar drinks in the U.S. For example, on weekend nights, approximately 6500 people cross back in to the U.S. between 12 a.m. and 4 a.m. after drinking in a bar or restaurant. Lange and Voas (2000) stated that, at least in California, youth escape the limits placed on drinking by crossing into Mexico where they can easily purchase alcohol and binge drink. In their study, over a one year period of time, they surveyed drivers and pedestrians on Wednesdays, Fridays and Saturdays who were crossing the Mexico-U.S. border in the northbound direction between 12 a.m. and 4 a.m. They asked participants to provide breath samples. Of the 5849 border crossers that were recruited, 87% completed the survey. They found that on weekend nights, pedestrians represented the highest concentration of drinkers, with more than 30% of them having a blood alcohol concentration (BAC) of .08% or greater. This finding was disconcerting since most of these pedestrians returned to parked vehicles on the U.S. side and drove or rode home. Therefore, the flow of young binge drinkers at the Mexico border is substantial, with this being only one of such binge-drinking locales where drinking age and cost of alcohol differs from one side of the border to the other.

AMETHYST INITIATIVE

Proponents of MLDA-21 have argued that raising the drinking age has saved lives (from reduced deaths from younger impaired drivers). However, others have argued that the MLDA-21 has merely pushed drinking in adolescents and college students into hiding, and subsequently drinking has become more extreme. The Amethyst Initiative was launched in July 2008 as a movement in the U.S. to debate the effectiveness of the 21 year old drinking age in the context of the significant binge drinking problem in high school and college students. The organization was founded by Dr. John McCardell, President Emeritus of Middlebury College (in Vermont) who has stated in various outlets that the MLDA-21 in the United States may be exacerbating the binge drinking problem on college campuses. For example, in the New York Times in 2004, he stated that "the 21-year-old drinking age is bad social policy and a terrible law" that has made the college drinking problem far worse. The Amethyst Initiative signers include 130 chancellors and presidents of various universities and colleges across the United States. Notable signatories include President Richard Brodhead (Duke University), President William Brody (Johns Hopkins University), President E. Gordon Gee (Ohio State University), President Lawrence S. Bacow (Tufts University), and Chancellor Robert Clarke (Vermont State Colleges).

Specific details about Amethyst Initiative can be found on their website (www.amethystinitiative.org). In an official statement, the organization detailed some of the reasons why MLDA-21 should be reconsidered. They argue that in the 24 years following the passing of the National Minimum Drinking Age Act, binge drinking on college campuses has escalated. A culture of dangerous and clandestine binge drinking has developed at off-campus locations since most college students are not of age to drink. The currently available alcohol education programs, with their abstinence approach (since that is the only legal option) have generally been ineffective in changing the behavior of students. Those students who choose to use fake identification to purchase alcohol are making ethical compromises that erode respect for the law. Proponents of a return to a lower drinking age (similar to that of Canada and Mexico) might be appropriate considering that U.S. adults under 21 years of age are deemed capable of voting, signing contracts, serving on juries, and enlisting in the military, but they are not mature enough to have a beer.

A lower drinking age might afford greater control over drinking. For example, a bartender at a campus bar can refuse to serve an intoxicated patron. Various supporting statements are listed on the Amethyst Initiative website, including a statement by Duke University President Richard H. Brodhead, which states,

> *"...It affects any school with an undergraduate population.*
> *Possessing and consuming alcoholic beverages is against the law under the age of 21, and we are all obliged to uphold the law. The current law has not prevented alcohol from being available and drinking is widespread at all American colleges and at younger ages as well. But at colleges and universities, the law does have other effects: it pushes drinking into hiding, heightening its risks, including risk from drunken driving, and it prevents us from addressing drinking with students as an issue of responsible choice..."*

It remains to be seen whether the Amethyst Initiative and the push for a lower drinking age will receive much traction and public support in the United States. However, culture clearly plays a critical role in binge drinking patterns and attitudes, and the current culture on many college campuses is a culture that embraces extreme binge drinking. Proponents of a lower drinking age in the United States have argued that the drinking age is unusually high compared to other countries around the world (see Figure 5-1).

Nevertheless, culture may matter more than alcohol control policies. Many eastern and northern European countries (i.e., the vodka belt countries) have cultures where binge drinking is the normative drinking pattern and drunkenness is a common outcome. For example, the United Kingdom, Ireland, and Russia all have age 18 as the legal drinking and purchasing age and binge drinking is common in those countries. A review of studies done in the United Kingdom (UK) from the past three decades demonstrated that binge drinking rates are extremely high in UK college students and these binge drinking rates may actually exceed the rates of their U.S. peers, despite a lower drinking age in the UK (Gill, 2002).

Meanwhile, Mediterranean countries such as Greece, Italy and France have drinking ages as low as 16 years with lenient enforcement and low rates of binge drinking. The traditional Mediterranean drinking culture does not generally condone either binge drinking or drunkenness (Heath, 1995, 1998, 2000; Room, 2001; Room and Makela, 2000). Clearly, a change in culture would help moderate high school and college binge drinking problems in North America. However, no-one has yet figured out how to make the U.S. more similar to the Mediterranean culture in their approach to alcohol.

Table 5.1. Minimum Age Limits and Standard Blood Alcohol Concentration Limits for Driving for various countries around the world

Country	Drinking Age: On premise	Purchasing Age: Off premise	Standard BAC (g%) limit for driving	Notes
Albania	None	None	.01	
Algeria	18	18	.01	
Argentina	18	18	.05	
Australia	18	18	.05	
Azerbaijan	18	18	0	
Belgium	16 (beer, wine); 18 (spirits)	None (beer, wine); 18 (spirits)	.05	
Bangladesh	Illegal	Illegal	0	For Muslims, alcohol consumption is illegal. However, foreigners may consume alcohol and it is served in hotels and restaurants.
Bolivia	18	18	.07	
Botswana	18	18	.08	
Brazil	18	18	.02	
Brunei	Illegal	Illegal	0	For Muslims, alcohol consumption is illegal. Non-Muslim residents and visitors may consume small amounts of alcohol.
Bulgaria	18	18	.05	
Cambodia	None	None	.05	
Canada	18/19	18/19	.08	18 in Alberta, Manitoba and Quebec; 19 in British Columbia, Saskatchewan, Ontario, Nova Scotia, New Brunswick, Prince Edward Island, Newfoundland and Labrador, Yukon, Northwest Territories and Nunavut
China	18	18	.05	
Columbia	18	18	0	
Costa Rica	18	18	.049	
Croatia	18	18	0	
Denmark	18	16	.05	
Ecuador	18	18	.07	
El Salvador	18	18	.05	
Estonia	18	18	.02	
Ethiopia	18	18	0	
Finland	18	18	.05	
France	16 (beer, wine); 18 (spirits)	16 (beer, wine); 18 (spirits	.05	
Georgia	16	16	.03	

Table 5.1. (Continued)

Country	Drinking Age: On premise	Purchasing Age: Off premise	Standard BAC (g%) limit for driving	Notes
Germany	16 (beer, wine); 18 (spirits)	16 (beer, wine); 18 (spirits)	.05	
Greece	17	None	.05	
Guatemala	18	18	.08	
Hungary	18	18	0	
Iceland	20	20	.05	
India	18 to 25 (depending on the state)	18 to 25 (depending on the state)	.03	Alcohol consumption is prohibited in the states of Gujarat, Manipur and Mizoram.
Iraq	18	18	0	
Iran	Illegal	Illegal	0	Alcohol can only be used for Jewish or Christian religious ceremonies
Ireland	18	18	.08	
Israel	18	18	.05	
Italy	16	16	.05	
Japan	20	20	.03	Alcohol vending machines are widely available.
Kenya	18	18	.08	
Kuwait	Illegal	Illegal	0	
Latvia	18	18	.049	
Libya	Illegal	Illegal	0	
Lithuania	18	18	.04	
Luxembourg	16	None	.08	
Malta	16	16	.08	
Mauritius	18	18	.05	
Mexico	18	18	.08	
Netherlands	16; 18 (spirits with ABV over 15%)	16; 18 (spirits with ABV over 15%)	.05	
New Zealand	18	18	.08	
Nicaragua	19	19	.08	
Norway	18; 20 (spirits 22% ABV or greater)	18; 20 (spirits 22% ABV or greater)	.02	
Panama	18	18	0	
Paraguay	20	20	.08	
Peru	18	18	.05	
Philippines	18	18	.05	
Russia	18	18	.03	
Saudi Arabia	Illegal	Illegal	0	
Singapore	18	18	.08	
Slovenia	18	18	.05	
South Africa	18	18	.05	
Spain	18	18	.05	
Sudan	Illegal	Illegal	0	

Country	Drinking Age: On premise	Purchasing Age: Off premise	Standard BAC (g%) limit for driving	Notes
Sweden	18	20; 18 (beer with ABV of 3.5% or less)	.02	
Switzerland	16/18 (beer, wine depending on region); 18 (spirits)	16/18 (beer, wine depending on region); 18 (spirits)	.05	
Thailand	18	18	.05	
Turkey	18	18	.05	
Turkmenistan	18	18	.03	
Uganda	18	18	.05	
United Kingdom	18 (see notes)	18	.08	Beer and wine can be purchased on premise at 16 if with a meal and when accompanied by an adult (at least 18 years or older).
United States	21	21	.08	
Uruguay	18	18	.08	
Venezuela	18	18	.05	
Zimbabwe	18	18	.08	

Data include the minimum age to legally purchase and consume alcohol. On-premise retail sale refers to the selling of alcohol for consumption at the site of sale (e.g., bar, pub, restaurant etc.). Off-premise retail sale refers to the selling of alcohol for consumption elsewhere and not at the point of sale (e.g., wine stores, supermarkets, gas stations, state monopoly stores, etc.). In most countries, the legal age to purchase alcohol is 18 but there are considerable variations and the degree to which these laws are enforced varies even within jurisdictions. In some countries, all alcoholic beverages are illegal. Sources include The World Health Organization (2004) and the International Center for Alcohol Policies (2008).

DRINKING AGE AND LIMIT FOR DRIVING UNDER ALCOHOL

Proponents of lowering the drinking age often cite the responsible drinking of countries like Italy as a reason to change MLDA-21 in the U.S. Table 5.1 does reveal that the drinking age is lower in many countries around the world compared to the United States. However, many countries with lower drinking ages also have corresponding lower blood alcohol limits for driving. For example, Italy allows individuals 16 years of age and older to drink alcohol and purchase alcohol. However, the Italian limit for driving is .05. Russia, which has a significant binge drinking problem, also has an 18-year-old legal limit for drinking. The Russian limit for driving is .03. As such, comparing alcohol control policies across various countries is complicated by the fact that multiple policies are often in place. Nonetheless, social factors, including culture, are extremely important variables to consider when interpreting causes of binge drinking and attempting to reduce the harms and hazards associated with this behavior.

CONCLUSION

Binge drinking in high school and college students is a social phenomenon. A variety of social and environmental factors are associated with binge drinking, such as college athletics and fraternity/sorority membership. Other environmental factors associated with binge drinking can include factors such as proximity of a U.S. border to Mexico (where the drinking age is 18 and drinks are less expensive). While the Minimum Legal Drinking Age of 21 in the United States has benefits to health and well-being, there are proponents (e.g., presidents and chancellors of U.S. universities that support the Amethyst Initiative) that argue that a 21 year old drinking age may be encouraging binge drinking in college students. Laws regulating the appropriate drinking age and the appropriate blood alcohol limit for driving differ around the world. It remains uncertain whether changes in alcohol policies in the United States could reduce the binge drinking problem in young people.

Chapter 6

NEUROCOGNITIVE AND HEALTH EFFECTS OF BINGE DRINKING IN ADOLESCENTS AND COLLEGE STUDENTS

Binge drinking in young people appears to be increasing in the United States, Britain, and in developing countries throughout the world (Naimi et al., 2003; Parry et al., 2002; Pincock, 2003). In South Africa, a recent survey reported that half of all male high school students were binge drinking by Grade 11 (Parry et al., 2002). A substantial body of research has documented the deleterious effects of heavy chronic drinking on health and well-being. Long-term abusive daily drinking can lead to numerous harmful effects on a variety of body systems. Liver damage, gastritis, pancreatitis, impaired reproductive functioning, impaired immune functioning, impaired cognitive functioning, and increased risk for a variety of cancers are among the health effects associated with chronic, heavy drinking (Bradley et al., 1993; Maisto et al., 2008). However, binge drinking is not like chronic drinking. Binge drinking is episodic in nature with many days of abstinence from alcohol as part of the pattern of drinking. If binge drinking rates are on the rise worldwide, it begs the question of whether the cognitive and health consequences of binge drinking are as serious as chronic drinking. If so, what are the implications of this behavior on general health and brain functioning?

BINGE DRINKING AND COGNITIVE FUNCTION

When alcohol is present in the body, several cognitive processes, including attention and memory, are impaired. These effects are reversible as soon as the blood alcohol concentration declines. However, chronic heavy drinking seems to impair these same cognitive functions in a way that leads to these processes remaining impaired even when the individual is sober. Alcoholism is often associated with brain damage and cognitive deficits. For heavy drinkers, the chronic effects of alcohol vary in severity, from mild to severe irreversible brain structural and functional damage (Maisto et al., 2008). Until recently, it was thought that only daily heavy chronic drinking for years led to such brain damage. More recent evidence suggests that even binge drinking, with its episodic nature, may be causing some damage to neurocognitive functioning.

Over the past few decades, there has been a slow accumulation of evidence, albeit controversial, that there is a negative association between the extent of social drinking and sober cognitive performance. The more frequently and more heavily a person drinks, the more poorly that person will often perform on a variety of cognitive tasks when tested in a sober state (Sher, 2006; Zeigler et al., 2005). This association has been demonstrated in both men and women, and decrements in performance are more evident in the older and/or heavier drinkers compared to the younger and/or lighter drinkers (Noble, 1983; Parsons and Nixon, 1998).

In alcoholics, morphological abnormalities in the frontal lobes of the brain have been reported (Moselhy et al., 2001). The frontal lobes are critical for impulse control, which explains why dysfunction in the frontal lobes may contribute to difficulty in recovering from alcoholism. If an individual has poor impulse control in general, he or she will have great difficulty controlling his or her drinking. Since this area of the brain seems to be affected in chronic heavy drinkers, researchers have been interested in determining whether frontal lobe function is normal or impaired in binge drinkers. A variety of neuropsychological tests have been developed to assess frontal lobe function, including tasks that measure the ability to inhibit a prepotent (i.e., instigated) response. Poor impulse control would be evidenced by an inability to withhold responses. Townshend and Duka (2005) used the Vigilance Task in the Gordon Diagnostic System to look at drinking status and frontal lobe performance (particularly impulsivity) in sober participants. In this vigilance task, participants repeatedly see three digits that are presented in random succession on a three column computer display. Participants are instructed to only concentrate on the digit in the middle column of the display. They are instructed to press a button every time a "1" is followed by a "9". Thus, the other two columns are only present as distraction. The researchers were most interested in when participants make errors of commission. Commission errors occur when a response is made to a "9" or a "1" when they are not in the exact "1" followed by "9" sequence. Thus, this task measures the subject's ability to inhibit responding under conditions that demand impulse control.

When Townshend and Duka (2005) compared the performance of binge and non-binge drinkers on this vigilance task, they observed that female binge drinkers made more commission errors compared to female non-binge drinkers. This difference was not replicated in the male binge and non-binge drinkers, and why this occurred is not known. However, despite the authors suggestion that vigilance task performance assesses impulsivity, the vigilance task is probably not a pure measure of one cognitive process since task performance not only requires impulse control, but also requires sustained attention. The increased impulsive errors made by the female binge drinkers do suggest impairment in inhibitory control, a cognitive process that is often attributed to frontal lobe activity (Townshend and Duka, 2005).

Binge drinking has been associated with impaired cognitive function in other domains. Weissenborn and Duka (2003) examined cognitive function in 95 social drinkers. They observed that the binge drinkers performed more poorly on a spatial working memory task and on a pattern recognition task compared to non-binge drinkers. This finding is similar to findings from other studies that have used brain imaging techniques to assess impairments in brain activity in binge drinkers. Functional magnetic resonance imaging (fMRI) studies can look at brain activity while a person is completing a cognitive task. Several fMRI studies have reported abnormalities in the brain's response to a working memory task in adolescents and young adults with alcohol use disorders (Sher, 2006). While an alcohol use disorder is a more severe problem than binge drinking, the finding that binge drinkers and alcohol dependent individuals both demonstrate working memory deficits is compelling evidence that the excessive use of alcohol may be causing some neural damage.

Hartley et al. (2004) also reported several cognitive deficits associated with binge drinking when binge and non-binge drinkers were compared on a variety of tasks that assess sustained attention, memory and executive function. These authors found that the binge drinkers made fewer correct responses on a test of sustained attention compared to non-binge drinkers. The binge drinkers also recalled fewer line drawings in a memory task compared to non-binge drinkers. Furthermore, the binge drinkers exhibited poor planning (a frontal lobe task that assessed executive function) on a test of planning compared to non-binge drinkers.

Decision making is another cognitive skill that has been of interest to researchers. In alcohol dependent individuals, disadvantageous decision making is observed compared with individuals without substance dependence (Bechara et al., 2001). Decision-making skills are especially important for high school and college students. Young adulthood is a time of change and educational and career development. Poor decision making skills at this age could lead to many problems that follow the individual well into adulthood. Goudriaan et al. (2007) measured behavioral decision making, using the Iowa Gambling Task (IGT) in binge and non-binge drinkers. In this task, participants play a card game on the computer in which they attempt to develop a winning strategy by selecting cards from advantageous versus disadvantageous decks. Participants learn, over the course of task, which decks are advantageous and which are disadvantageous. Disadvantageous choice patterns indicate poor decision making. Goudriaan et al. (2007) observed that disadvantageous decision making was related to binge-drinking patterns. Since decision making skills rely on prefrontal lobe functioning, some impairment in this region may be occurring. In humans, the prefrontal lobes mature through adolescence and into early adulthood (Casey et al., 2000). Therefore, binge drinking may be harming a region of the brain that is still developing. The long term implications of this are still unknown.

Zeigler et al. (2005) reviewed numerous studies on the neurocognitive effects of alcohol on adolescents and college students. They reported that underage drinkers were at elevated risk of neurodegeneration (particularly in those regions of the brain used for tasks like learning and memory). If underage alcohol consumption is associated with brain damage and neurocognitive deficits, there are significant implications for learning and intellectual development. Impaired intellectual progress will continue to affect a person well into adulthood. The extent of neurodegeneration and whether drinking causes any permanent damage that cannot be reversed with abstinence from alcohol is still unknown.

It is important to note, however, that the association between binge drinking and cognitive dysfunction is new and controversial. Many null results have been reported, although some of those null results might be due to the small sample sizes used (Parson and Nixon, 1998; Rose and Grunsell, 2008). For example, Rose and Grunsell (2008) compared binge and non-binge drinkers on two tasks that measured impulsivity (Two Choice and Time Estimation). They found that binge drinking status was not associated with impulsivity task performance. However, their study had a small sample size of 20 participants, 10 of which were binge drinkers. Had they recruited more participants, would they have found group differences in impulsivity like Townshend and Duka (2005) did using a larger sample size? Another concern with this literature is that the association between binge drinking and cognitive dysfunction is a correlation which makes causal statements difficult. An alternative explanation of the association between heavy social drinking and cognitive dysfunction is that individuals with poorer cognitive functioning to start with are more attracted to use alcohol (and not that the binge drinking led to a decrement in cognitive functioning). Until someone completes a prospective study which compares adolescents before and after they have indulged in binge drinking behavior, the clear distinction of what cognitive impairment preceded and what followed as a result of a pattern of binge drinking is impossible to ascertain.

Even though causal statements are currently difficult to make, the association between binge drinking and cognitive dysfunction leads to an important question. Why would binge alcohol consumption lead to damage in neurocognitive functioning? One proposed answer is that this outcome may be due to the alternating pattern of drinking to intoxication followed by total abstinence being quite damaging to the brain. Studies have shown that chronic administration of alcohol leads to upregulation of N-methyl-D-Aspartate (NMDA) and calcium receptors and increased release of glucocorticoids in the brain. It is known that NMDA-mediated mechanisms and glucocorticoid actions on the hippocampus are associated with brain damage. Therefore, every time alcohol is withdrawn, it may make the brain more vulnerable to damage from these mechanisms (Hunt, 1993). Since binge drinking by nature involves swings from heavy consumption to periods of total withdrawal from alcohol, this damage might be more likely to occur.

Preliminary evidence that withdrawal from alcohol may be an important component of neurocognitive damage from binge drinking was provided by McKinney and Coyle (2004). These authors investigated the effects of drinking on next day memory and psychomotor performance in social drinkers. In their study, participants went about their normal drinking routine (usually to intoxication) with the only limitation being that drinking had to be restricted to the period between 22:00 and 02:00 hours on the night before testing. Participants were also tested when they had not been drinking the previous night. The authors reported that the morning after alcohol was consumed, performance on memory tasks and on a psychomotor task were impaired, despite the fact that blood alcohol levels were zero at the time of testing.

BINGE DRINKING AND MOOD

Adolescent and college-aged binge drinkers often associate binge drinking with parties, social gatherings and fun. However, when researchers have measured mood states, they have actually found that the general mood of sober volunteers differs depending on their drinking habits. Binge drinkers often self-report having less positive mood than non-binge drinkers. Townshend and Duka (2005) measured mood in a sample of college age social drinkers using the profile of mood states questionnaires. This scale consists of 72 mood adjectives which participants are instructed to rate on a five point scale. Responses on this scale are used to assess a variety of mood states including anger, anxiety, fatigue, depression, vigor, confusion, friendliness and elation. The authors found that binge drinkers had lower scores for the positive moods. Because the time of last drink was recorded, it could be argued that low current sober mood might reflect withdrawal from alcohol in the binge drinkers. It turns out that there was no relationship between current positive mood and time of last drink. Lower overall mood in binge drinkers is similar to findings reported with alcoholics. Increased anxiety and negative emotional sensitivity has been reported in alcohol dependent individuals (Duka et al., 2002). However, the cause and effect relationship between alcohol abuse and low mood is unclear. It could be that increased anxiety advances the progression to alcohol dependence just as the heavy use of alcohol decreases mood.

Also important in the discussion of mood and binge drinking is the possibility that low mood could contribute to the neurocognitive deficits seen in binge drinkers. Low mood is associated with low energy and low motivation to perform well on various laboratory tasks. However, not all studies have found that binge drinkers display low mood. Hartley et al. (2004) actually observed the opposite pattern, with binge drinkers having significantly lower ratings of depression and anxiety at the time of testing compared to individuals who didn't drink. However, binge-drinking participants in the Hartley et al. study still performed worse on a variety of cognitive tasks compared to the nondrinkers. As such, mood ratings were not associated with neurocognitive performance in the Hartley et al. study, suggesting that not all poor cognitive performance in binge drinkers is driven by mood levels.

BINGE DRINKING AND SUICIDE RISK AND RISK FOR AGGRESSION

Demonstrations that binge drinking is associated with poor neurocognitive performance and lowered mood (at least in some cases) has led investigators to examine the implications that this pattern may have on other behaviors, including suicidality. Both chronic and acute alcohol use has long been identified as a risk factor for suicidal behavior. In one review, O'Connell and Lawlor (2005) observed that the cognitive effects of binge drinking, including deficits in attention, prospective cognition, autobiographical memory and disinhibition, combined with emotional mechanisms such as dysphoria, depression and aggression may lead to suicides. Similarly, Parrott and Giancola (2006) investigated the acute effects of alcohol on aggressive behavior in the laboratory. They found that alcohol increased aggression, but only in those men who had a history of heavy episodic (binge) drinking.

The purported neurocognitive damage and other health effects caused by binge drinking may be leading to deaths both directly and indirectly. For example, Chenet et al. (2001) studied increased mortality rates in Lithuania in the 1990s. During this same time period, binge drinking also escalated. They found that daily variations of deaths were associated with binge drinking. Even though an individual may have died from cardiovascular disease, he or she had often been binge drinking during the previous day. Thus, binge drinking impacts overall health and may be acting as a catalyst for pathophysiological events by increasing blood pressure, cardiac rhythm and coagulability. Binge drinking has also been associated with death rates in studies completed in Russia (Pridemore, 2004).

These findings have led other investigators to note that the relationship between alcohol and cardiovascular health may be a razor-sharp double-edged sword. There appears to be a J-shaped association between alcohol intake and a variety of adverse health outcomes, including coronary heart disease, hypertension, congestive heart failures, stroke, diabetes, and dementia.

While light drinking (up to 1 drink daily for women and up to 2 drinks daily for men) is associated with cardioprotective benefits, any amount of alcohol consumption beyond that level results in proportional weakening of outcomes (hence the J shape association).

Binge drinking, even among otherwise light drinkers, increases cardiometabolic risk, cardiovascular events and mortality (Fan et al., 2008; O'Keefe et al., 2007). Similar concerns with binge drinking also apply to mental illness. Haynes et al. (2008) followed patients who initially presented with anxiety and depression over an 18 month period of time. The authors observed that compared with the non-binge drinking group, binge drinkers were less likely to have recovered from this common mental disorder at follow-up.

BINGE DRINKING AND DIFFERENT REACTIONS TO ALCOHOL STIMULI

In current theories of addiction, stimuli associated with drug use change from being neutral in valence to having motivational salience and capturing attention. In other words, the sight of a bottle of beer elicits craving in an alcoholic even though this stimulus is relatively neutral in value for someone who doesn't drink. Attentional bias is one technique researchers have used to observe this change in reaction to stimuli in addiction. For example, previous research has shown an attentional bias toward drug-related stimuli in opiate addicts.

Townshend and Duka (2001) examined whether heavy social drinkers also have an attentional bias toward alcohol-related stimuli compared with a group of occasional social drinkers. To measure attentional bias, they used a dot-probe detection task. In this task, pictures and words pairs were visually presented on a computer screen. Following the word/picture pair, a dot on the computer screen replaced one of the items. Attentional bias was measured by the reaction time to respond to the dot.

They observed that heavy social drinkers were faster to respond than lighter drinkers to a dot probe when it followed an alcohol picture. In other words, the heavy social drinkers had a greater attentional bias toward the alcohol-related stimuli (i.e., their attention was captured more readily by alcohol-related stimuli).

Other paradigms have been used to assess attentional bias, such as Stroop. In the classic Stroop Task, participants are instructed to name the ink color of stimuli and words. In the Stroop condition, participants are given an actual color word that is typed in another color (e.g., the word red is typed in blue ink and the participant must state 'blue'). Typically, participants have difficulty in overriding their natural tendency to read words and have difficulty responding quickly and accurately in this Stroop condition. In a modified Stroop task, researchers use other words besides color words that might be distracting to color naming performance.

For example, alcohol words such as 'keg' or 'beer' may also be distracting for alcoholics, indicating greater attentional bias to alcohol-related stimuli. It has also been shown that greater alcohol attentional bias is present in heavier social drinkers compared to lighter social drinking controls (Bruce and Jones, 2004). Changes in attentional bias toward alcohol-related stimuli may be a starting point along a path toward addiction. However, studies are needed to determine whether level of attentional bias is associated with later development of alcohol abuse and dependence.

BINGE DRINKING AND DIFFERENT REACTION TO THE ACUTE EFFECTS OF ALCOHOL

Binge drinking may not only change reactions to alcohol stimuli in the environment but also the reaction to the consumption of alcohol itself. Among social drinkers, heavier drinkers often report greater stimulant-like and fewer sedative-like effects and aversive subjective effects after alcohol consumption compared to lighter drinkers (Brumback et al., 2000; Evans and Levin, 2004; Holdstock et al., 2000; Marczinski et al., 2007, 2008). These differential subjective responses to the alcohol are at odds with the fact that alcohol similarly impairs various cognitive processes in both groups of drinkers. These findings of differential subjective reactions to alcohol are observed even though breath alcohol levels are equivalent between the two groups of drinkers. It is unclear why binge drinkers report different effects of alcohol. One possibility is that heavy binge social drinkers judge interoceptive cues from alcohol intoxication to be less severe because they are more familiar with them (Brumback et al., 2007). Binge drinkers could also be developing greater tolerance to the subjective effects of alcohol because they drink more heavily and drink more often than lighter social drinkers (Kalant et al., 1971; Marczinski and Fillmore, in press). More importantly, these differential subjective reactions to alcohol in binge drinkers have clinical relevance and public safety implications. For example, the subjective reactions to alcohol in binge drinkers may lead the binge drinkers to think they can drive, even when they shouldn't (Marczinski et al., 2007, 2008).

EARLY-ONSET ALCOHOL DRINKING

Individuals who start drinking before the age of 14 years are 12 times more likely to be injured in accidents while under the influence of alcohol than those who start drinking after age 21 (Hingson et al., 2000). Early-onset drinking is also significantly associated with lifetime alcohol problems, such as alcohol dependence (Chou and Pickering, 1992). While few studies have examined the impact of early-onset alcohol consumption on neurocognitive performance, early evidence suggests that earlier drinking is more damaging to cognitive performance. Goudriaan et al. (2007) found that in college students, heavy alcohol use at an earlier time (i.e., in high school) was more strongly related to diminished decision-making skills in sober participants compared to those participants who had begun heavy drinking more recently (i.e., during college). If the brain is still developing during adolescence (Casey et al., 2000), heavy drinking may be more harmful to the brain during that time period compared to heavy drinking in a fully developed brain. However, prospective studies are needed. It could be that diminished decision-making skills are already present before the onset of binge drinking and diminished decision-making skills promote early-onset binge drinking. Thus, separating cause from effect is nearly impossible until a study is designed that examines cognitive performance of high school students before and after initiation of binge drinking.

CONCLUSION

Binge drinking may lead to deleterious effects on health, particularly on brain functioning. A variety of studies have noted that binge drinkers display poorer performance on a variety of neurocognitive tasks compared to non-binge drinkers. The brain continues to mature into young adulthood. Thus, the heavy consumption of alcohol during brain development may cause more damage than would be observed with a fully developed brain. However, the cause and effect relationship between binge drinking and neurocognitive deficits is still tentative. Prospective studies are needed to delineate whether binge drinking leads to cognitive deficits or whether various cognitive deficits lead to poor decision making and impulsivity, which in turn lead to binge drinking. Binge drinking has also been associated with greater attentional bias toward alcohol-related stimuli and more positive subjective reactions to the acute effects of alcohol. These findings suggest that binge social drinking, while separate from alcoholism, may be a milestone along a path toward more serious problems with alcohol abuse and dependence.

Chapter 7

ANIMAL MODELS OF ADOLESCENT AND ADULT BINGE ALCOHOL USE

There is considerable evidence that the chronic abuse of alcohol, as often seen in an alcoholic, can result in significant brain damage, such as overall reduction in brain volume (Agartz et al., 1999; Pfefferbaum et al., 1992). However, researchers have been unable to consistently correlate lifetime consumption of alcohol with neuropathological changes (Dent et al., 1997). This led several groups of researchers independently to suggest that the drinking pattern (binge versus chronic) matters more than the amount of alcohol consumed when one is looking for brain damage. Binge drinking that leads to intoxication may be the key factor in observed brain neuropathology (Agartz et al., 1999; Crews, 1999; Fadda and Rossetti, 1998; Hunt, 1993).

In the previous chapter, we provided evidence from behavioral tasks that demonstrated that binge drinking in adolescents and young adults may lead to deleterious effects on brain functioning. College student binge drinkers display poorer performance on a variety of neurocognitive tasks compared to non-binge drinkers (Goudriaan et al., 2007; Hartley et al., 2004; Townshend and Duka, 2005; Weissenborn and Duka, 2003; Zeigler et al., 2005). This outcome may not be surprising considering that the brain continues to mature into young adulthood (Chambers et al., 2003). Brain constituents that actively develop during adolescence include the prefrontal cortex, limb system areas (e.g., hippocampus, amygdala), and white matter myelin (Clark et al., 2008). Therefore, heavy consumption of alcohol during adolescent neurodevelopment may cause more damage than what might be observed in a fully developed (i.e., adult) brain (Chambers et al., 2003; Clarke et al., 2008). Developmental delays or deficits in various structures and functions of the brain may also underlie the accelerated alcohol use observed in adolescence (Clark et al., 2008). However, there are no human prospective studies available that have investigated cognitive performance in adolescents before and after the onset of binge drinking. Thus, it is impossible to determine from the available human studies whether neurocognitive changes are a cause or consequence of binge drinking. It seems just as reasonable to hypothesize that poor neurocognitive performance may lead an individual to begin binge drinking as it is to hypothesize that binge drinking leads to neurocognitive impairment. Furthermore, both processes could be at play (i.e., poor neurocognition leads to binge drinking which leads to further deterioration in neurocognition). Fortunately, animal studies on binge drinking afford greater control and thus help in our interpretation of whether binge drinking leads to brain damage.

ANIMAL MODELS OF BINGE DRINKING

Most animal models of binge drinking involve the administration of high doses of alcohol over short durations to rodents. The advantage of using animals as subjects, versus studying high school and college students who already binge drink, is that the animals that start an experiment are naïve to alcohol. Furthermore, the various social factors that impact human drinking are not present in a laboratory study of binge drinking using rats. Control is achieved in animal studies as animals are randomly assigned to a condition where they will 'binge drink' or another condition where they will not (or where they consume alcohol, but in a more moderate fashion). Following drinking, animals might be tested on behavioral tasks and/or researchers might examine the brains of the animals post-mortem to look for evidence of damage. The heightened control present in animal experiments allows researchers to make stronger causative statements about the effects of binge drinking on brain functioning.

To model the pattern of alcohol consumption that is observed in human binge drinking using rodents, the protocol often involves the heavy administration of alcohol over a short time period with periods of complete withdrawal from alcohol (similar to a college student drinking excessively on the weekend followed by abstinence during the week). For example, Crews and colleagues have modeled binge drinking by examining the effects of a 2 to 4 day binge ethanol treatment in rats. With this binge protocol, rats were administered 15% ethanol intragastrically 4 times per day for 2 to 4 days. This dosing regimen resulted in the consumption of approximately 10 g/kg per day, a very high amount of alcohol that maintains intoxication but still minimizes animal mortality (Crews et al., 2000; Obernier et al., 2002). An alternative method used by Stephens and colleagues is to provide rats with an alcohol-containing diet as their sole source of nutrition for 24 days, with several interruptions of withdrawal periods (Stephens et al., 2001, 2005; Stephens and Duka, in press). While both methods might produce blood alcohol concentrations that are quite high, they are not out of range of that observed with adolescents and young adults who are treated for alcohol intoxication in a hospital emergency department. Thus, these models can provide important information about what extremely high doses of alcohol followed by withdrawal periods of no drinking do to the body. Results from these animal studies have informed us how a pattern of alcohol use alternating from the extreme of alcohol intoxication to a period of abstinence may cause significant brain damage (Crews et al., 2000; Stephens and Duka, in press).

The results from these animal studies have provided us with extremely compelling evidence that binge drinking is harmful. Even small amounts of binge drinking can cause damage. For example, one study exposed rats to a two or four day binge of alcohol and then the rats were sacrificed immediately or 3 days after withdrawal to examine their brains for damage. After only a two day binge, significant damage was observed in the olfactory bulb of the rat brains. This pattern of alcohol consumption is not uncommon to a college student who goes away for spring break and remains intoxicated over a two day period of time. For those animals exposed to a four day binge, further damage was observed in various parts of the brain. In addition, the necrotic degeneration that was observed in the brains after 2 days of exposure and the increased damage observed in the brains after 4 days of the binge was not increased during withdrawal. Therefore, it is the binge drinking that induces the brain damage and not the alcohol withdrawal (Obernier et al., 2002). Similar studies that have extended the binge experience for rats have found that the damage escalates as exposure to intoxicating levels of alcohol increases. Adolescent rats exposed to intermittent alcohol intoxication over a two week period of time exhibited increased cell death in the neocortex, hippocampus and cerebellum of the brain (Pascual et al., 2007). Therefore, a high school or college student who only binge drinks occasionally may not be immune to brain damage, despite the fact that they drink only occasionally. Only one extreme episode of intoxication may be sufficient to cause damage. More frequent episodes of intoxication are certain to cause more widespread damage. Whether this damage can be reversed is unknown.

BINGE DRINKING AND THE PREFRONTAL CORTEX

The outer covering of the brain is called the cerebral cortex; this region is especially prominent in humans (see the website www.brainmuseum.org for pictures of brains from various species). On the anterior (forward) section of the cerebral cortex is the prefrontal cortex (approximately where your forehead is located). This part of the brain is important for various aspects of decision making, planning and impulse control (Kalat, 2008). The prefrontal cortex is essential when you decide to pass up an immediate reward for a later reward (Frank and Claus, 2006). For example, when you chose to pass up going to a bar with friends because you need to go home and study to get a good grade on a test, you are relying on the prefrontal cortex for this decision. People who have prefrontal cortex damage make impulsive decisions (Anderson et al., 1999; Damasio, 1999). Documented as early as the year 1848, damage to the ventromedial prefrontal cortex causes pervasive motivational impulsivity associated with emotional (affective) instability. Thus, a person with prefrontal cortex damage will make poor decisions, have inability to plan, may be highly emotional and demonstrate indifference to social cues (Damasio et al., 1994).

As described in the previous chapter, prefrontal cortex function has been compared in college student binge and non-binge drinkers who were matched for age and IQ. Binge drinkers are more impaired on a vigilance task that requires the participant to withhold a prepotent response. The increased impulsivity in the binge drinkers is suggestive of a lack of inhibitory control from the frontal lobes and indirectly suggests that binge drinking may be damaging the prefrontal cortex (Townshend and Duka, 2005). In humans, age at which binge drinking started factors into the impairment observed, with earlier drinking associated with greater impairment. Initiation of heavy alcohol consumption in adolescence is associated with this impairment in cognitive function (Brown et al., 2000) and early exposure to binge drinking is associated with frontal lobe damage (Crews et al., 2007). However, the human literature has differing views on how to attribute causation. It has also been suggested that prefrontal dysfunction is a predisposing factor (and not a consequence) to binge drinking. Evidence for this view comes from studies of young adult social drinkers where a relationship was found between impaired executive function (a cognitive process that relies on the frontal lobes) and both the frequency of drinking to 'get high' and 'get drunk' (Deckel et al., 1995) and the severity of drinking consequences (Giancola et al., 1996). As such, animal studies can provide important clues about whether prefrontal cortex dysfunction is a cause or consequence of binge drinking.

To measure impulse control (i.e., prefrontal cortex functioning) in an animal study, the negative pattern task has been used. In this task, rats are required to initiate a response when either a light or a tone stimulus is presented, but to inhibit the response when both stimuli were presented simultaneously (Bussey et al., 2000). Deficits in performing the task are attributed to an inability to withhold responding, a process that uses the frontal cortex. Rats exposed to alcohol in a binge fashion perform poorly on this task in that they are impaired in suppressing prepotent responses (the rat responds when both the tone and light are presented simultaneously, even though it should not) suggesting that binge drinking may alter frontal cortical function (Borlikova et al., 2006).

In other animal studies, the effect of binge alcohol exposure on the prefrontal cortex has been more directly examined. Crews et al. (2000) exposed young adolescent rats (approximately 35 days old) to binge alcohol consumption and then sacrificed the animals to examine their brains. The authors observed increased levels of amino cupric silver staining (an indicator of cell death) in frontal areas in those binge animals. The findings from these animal studies are consistent with those findings with human binge drinkers and alcoholics who show impaired performance on tasks sensitive to dysfunction of the prefrontal cortex (Duka et al., 2003, 2004; Weissenborn and Duka, 2003; Townshend and Duka, 2005).

AMYGDALA

Another structure of the brain, the amygdala, has been a focus of attention for binge drinking researchers. The amygdala is a subcortical limbic structure found deep within the temporal lobe (think of the part of the brain that is above the ears and then moving inwards toward the center of the brain). This structure is critical for certain aspects of experiencing strong emotion and recalling highly emotional memories (Kalat, 2008). For example, individuals who experienced first-hand the terrorist attacks in downtown Manhattan New York City on September 11, 2001, have very strong memories of that experience. Researchers have brought these individuals into a laboratory and had them recall those 9/11 memories while imaging their brain using a technique called functional magnetic resonance imaging (fMRI). Recall of traumatic 9/11 memories involves selective activation in the amygdala, which is not observed when other control events are recalled. Furthermore, other individuals who were in New York City but not personally involved did not shown this selective amygdala activation (Sharot et al., 2007). Thus, the amygdala is an important brain structure involved in the emotional modulation of memory.

There is evidence for altered emotional reactivity in human binge drinkers, indicating that there may be a change in amygdala activity. For example, college binge drinkers have self-reported lower positive mood, using the profile of mood states questionnaire, compared to their more moderate drinking peers (Townshend and Duka, 2005).

To examine the relationship between binge drinking and amygdala activity, animal studies have relied on models that train an animal to acquire a conditional emotional response. For example, a rat will be trained that a presentation of a tone predicts a mild foot shock. As such, a control animal quickly learns to avoid the tone. This conditioned (i.e., learned) emotional response is processed within the amygdala (Seldon et al., 1991).

When animals were exposed to high doses of alcohol followed by many repeated withdrawals (i.e., binge alcohol consumption), researchers observed that binge drinking rats were slow to acquire this conditioned emotional response, even several weeks after cessation of the final alcohol treatment (Stephens et al., 2001). Interestingly, if the training of the conditioned emotional response task occurred before the binge alcohol exposure, these rats were not impaired in expressing the appropriate reaction. This suggests that the binge drinking leads to impairments in learning this tone and shock association, and not that the binge animals have a blunted fear response (Ripley et al., 2003). Binge drinking appeared to impair the learning of new emotional associations, but if those associations were learned prior to binge drinking, there was no impairment in the expression of the appropriate response (i.e., trying to escape the foot shock when the tone is sounded).

In an interesting extension to these findings, researchers have also demonstrated that the binge administration of alcohol is worse than the chronic administration of alcohol, at least for learned associations that require amygdala activity. Rats that were exposed to the binge administration of alcohol were compared to rats that were exposed to continuous alcohol exposure (in an equivalent amount). It was observed that binge rats were slower to learn the tone and foot shock association compared to the chronic drinking rats (Ripley et al., 2003; Stephens et al., 2001). Findings such as these have important implications to understanding the harms and hazards associated with binge drinking.

Many binge drinkers argue that their consumption patterns are harmless because they are not dependent on alcohol since they spend several days a week not drinking anything. However, it is the high level of intoxication on the days that a binge drinker does drink that may be causing deleterious alterations in brain functioning.

RELATIONSHIP BETWEEN PREFRONTAL CORTEX AND AMYGDALA

Human imaging studies have indicated that there is relationship between the prefrontal cortex and the amygdala (the two brain areas that researchers have found are altered in activity with binge drinking). The activity in the prefrontal cortex is inversely correlated with activity in the amygdala, suggesting that the prefrontal cortex may suppress amygdala-mediated responses (Hariri et al., 2000). As such, it has been speculated that binge use of alcohol followed by repeated episodes of withdrawal may impair prefrontal cortex function, and a consequence of this may be the predisposition to recall aversive experiences that are normally suppressed (Stephens et al., 2005).

Similarly, the loss of control of drug taking in addicts has been attributed to the loss of the ability of the prefrontal cortex to inhibit behaviors mediated by subcortical systems (such as the amygdala). The prefrontal cortex controls the executive functions, such as the ability to plan and to inhibit behaviors, which are cognitive processes that are critical for controlling excessive consumption of alcohol or drugs (Volkow et al., 2003). While highly speculative at this point, this brain pathway change with binge drinking may be setting up the precursors for more serious alcohol dependence problems.

MECHANISMS UNDERLYING THE EFFECTS OF
BINGE CONSUMPTION ON THE BRAIN

The effects of alcohol on the brain still remain a complicated puzzle to solve, particularly in the case of binge alcohol use. At least for the acute effects of alcohol, several neurotransmitter changes are understood. Acute alcohol treatment is associated with facilitation of GABAergic inhibitory mechanisms in the brain (Samson and Harris, 1992; Roberto et al., 2004). Alcohol also acts as an antagonist of glutamatergic NMDA receptors, meaning that alcohol blocks these receptors (Samson and Harris, 1992). When alcohol exposure is chronic, the transmission in the glutamatergic systems is facilitated via increased NMDA receptor sensitivity (Roberto et al., 2004) and increased glutamate turnover (Dahchour and De Witte, 1999). This is to compensate for the two major actions of alcohol (facilitation of GABA and antagonism of glutamate). These changes result in partial tolerance to the sedating effects of alcohol.

When alcohol is removed from the body, the glutamatergic system continues to be overactive while NMDA receptor function remains elevated for a period of time (Dahchour and De Witte, 1999; Roberto et al., 2004). However, since there is no longer any alcohol present, this overactivity is no longer balanced by alcohol's effects on GABAergic systems. Evidence is accumulating to illustrate that when alcohol is present and then absent repeatedly, this facilitates glutamatergic synaptic transmission in the amygdala (Floyd et al., 2003; Lack et al., 2007; Roberto et al., 2006), meaning that the amygdala is generally in an overactive state, even if the individual is not currently drinking. As such, binge drinking may have carryover effects of neurotransmitter changes to later sober days.

BINGE DRINKING AND THE DEVELOPING BRAIN

The brain continues to change and mature throughout childhood and adolescence. Particularly during adolescence, neurodevelopment involves changes in brain organization and function characterized by a greater relative influence of motivational substrates even while inhibitory substrates (i.e., impulse control) are still immature and developing. The increased motivational drive for new experiences coupled with this immature inhibitory control system is what predisposes a teenager to perform impulsive actions and risky behaviors (including experimenting with alcohol and other drugs) (Chambers et al., 2003). Unfortunately, there couldn't be a worse time than adolescence to ingest these substances. There are important changes occurring in the brain during those years that make alcohol more harmful to a developing brain compared to a fully developed brain (Spear and Varlinskaya, 2005).

To illustrate this problem, recent animal studies have compared the effects of binge alcohol consumption on subsequent brain damage in both adolescent and adult rats. Crews et al. (2000) compared rats that were 35 days old (juveniles who are analogous in age to a human adolescent) to rats that were 80 to 90 days old (adults). Following binge administration of alcohol over a 4 day period, the authors found that significant brain damage was found in both juvenile and adult rats (e.g., damage in the olfactory bulbs). However, the juvenile brains also exhibited damage beyond that observed with the adults. The juveniles had damage in the frontal cortical olfactory and anterior piriform and perirhinal cortical regions of the brain in addition to the olfactory bulb damage. These findings illustrate that the young adolescent brain may be more sensitive to alcohol-induced brain damage compared to an adult brain that has completed neurodevelopment. Many other studies have also noted the differential reactions of juvenile and adult rats to alcohol. Adolescent rats are less sensitive to the sedative action of alcohol, as measured by loss of righting reflex (the ability to stand back up if fallen over) (Little et al., 1996). However, adolescent rats show greater hypothermic response to alcohol (Swartzwelder et al., 1998) and display greater memory impairment following alcohol administration (Markwiese et al., 1998) compared to adult rats.

During adolescent neurodevelopment, one area of the brain that still is changing dramatically is the prefrontal cortex, an area that has not yet maximized various cognitive functions, including its ability to inhibit impulses. For example, performance on a variety of laboratory tasks that measure prefrontal cortex function, including working memory, complex problem solving, abstract thinking, sustained logical thinking, and the ability to inhibit psychomotor responses, improves dramatically through childhood and peaks in late adolescence (Chambers et al., 2003). Similarly, laboratory examinations of developmental changes in brain anatomy and function correspond temporally to reported changes in cognitive function. For example, the ratio of lateral ventricle to brain volume remains relatively constant during childhood (ages 6 to 12), but then increases steadily during adolescence (ages 12 to 18) (Giedd et al., 1996). Similarly, between the ages of 4 and 17, there is a progressive increase in white matter density in the frontal cortex. White matter reflects the presence of myelination of neurons. Myelinated axons of neurons increase the efficiency of action potential propagation, indicating that this change from gray matter over to more white matter means that the frontal cortex is now working effectively and efficiently by the end of adolescence (Jernigan et al., 1991; Paus et al., 1999). This developmental change of increased white matter in the prefrontal cortex as adolescents mature into adults is not unique to humans and has also been observed in rats (van Eden et al., 1990).

Other areas of the brain are also changing during adolescence. Maturational changes occur during puberty in the hippocampus (a region important for spatial memory) and the hypothalamus (a region important for motivated behaviors including hunger, thirst, temperature regulation, and sex drive) (Choi et al., 1997; Choi and Kellogg, 1992; Dumas and Foster, 1998; Wolfer and Lipp, 1995). Thus, the developmental changes that are occurring in the adolescent brain can be altered by alcohol in ways that are different from those in adults. However, although the maturation is progressive, it is far from uniform in speed and timing and individual differences are the rule, rather than the exception. For many of these maturational processes in the brain, there are periods of rapid transition, reorganization, and growth spurts that alternate with periods of consolidation and quiescence (Moss et al., 2008). Since development does not progress linearly, this makes understanding the effects of binge drinking on the brain extremely challenging. To add to the complexity, males and females differ in developmental changes in gray and white matter density. These sex differences in the brain may be under the regulatory control of sex hormones that surge during pubertal development (Moss et al., 2008). Despite all of these complications (nonuniform progression of development and sex differences), it is becoming well understood that adolescent exposure to alcohol may influence ongoing neural maturation and then later neural, cognitive, and behavioral functioning, including later sensitivity to and propensity to use alcohol (Spear and Varlinskaya, 2005). The results from animal studies clearly illustrate that binge drinking is more damaging to an adolescent brain than an adult brain and that is why strong efforts are needed to limit this behavior in our high school and college students.

CONCLUSION

Far from innocuous, binge drinking may cause significant brain damage in the user. In human binge drinkers and in animal models of binge patterns of alcohol consumption, there is behavioral and neurological evidence for altered functioning in many areas of the brain, including the prefrontal cortex and the amygdala. The prefrontal cortex is critical for impulse control and the amygdala is critical for emotional processing. Results from animal studies have illustrated that binge alcohol use can cause brain damage in these regions (e.g., cell death in the prefrontal cortex). Since neurodevelopment continues until late adolescence, binge drinking may be more damaging to an adolescent brain than a fully-developed adult brain. In human studies, the simultaneous development of higher brain areas and their interconnections, advancement in skill in behavioral regulation, and acceleration of alcohol involvement in adolescence makes the effects of binge drinking on the adolescent brain a complicated puzzle to solve. However, the increased control afforded by animal studies has revealed that binge drinking specifically causes brain damage, with greater damage seen in a developing brain. While the exact mechanisms of why binge alcohol use causes brain damage still need to be elucidated, it is clear from animal studies that this pattern of drinking causes serious harm.

Chapter 8

BINGE DRINKING, IMPAIRED DRIVING AND OTHER HEALTH RISK BEHAVIORS

While drinking to get drunk causes numerous direct risks to health and well-being, there is also the serious concern that binge drinking leads to many high risk behaviors that might not be engaged in when sober. Driving while intoxicated is one excellent example. Most high school and college students know that this behavior is highly dangerous and also illegal. However, many of them will think nothing of getting in a car and driving home when intoxicated or be a passenger in a car with someone who has also been drinking. Findings from one of the largest population-based studies of the drinking patterns of alcohol-impaired drivers in the United States demonstrated that binge drinkers accounted for 84% of alcohol-impaired drivers and 89% of alcohol-impaired driving episodes (Flowers et al., 2008). This may differ from the general public's preconceived notions that impaired drivers are alcoholics. More importantly, alcohol impaired driving has been found to be common for binge drinkers across educational and income categories. This finding differs from the usual inverse relationship obtained between income/education and other important health risk behaviors (e.g., smoking, unprotected sex). A binge drinker with a high annual income (i.e., > $75,000) and a college degree is just as likely to drive drunk as an individual with a low annual income and little education (Flowers et al., 2008). There is considerable drinking and driving done by high school students, despite extensive education campaigns that explain the dangers of combining these behaviors (Williams et al., 1986).

Other high risk behaviors performed while drinking can lead to serious injuries and deaths. One recent example is the death of Miami of Ohio college student, Beth Speidel, age 19. She was killed when she was struck by a freight train following a night of drinking on April 14, 2007. According to the coroner, Beth had a blood-alcohol level more than twice the Ohio state's legal limit of .08 g%. Her impaired judgment due to alcohol intoxication probably led to her risky behavior of walking too close to train tracks and a moving train on her way home from a bar. The tragedy of this death is not limited to her alone. In this case, several fellow Miami of Ohio University students were charged with alcohol offenses in connection with Beth's death. Danielle Davis, 20, Kathleen Byrne, 19, and Kristina Sicker, 20, were charged as was Maureen Grady, 20, who was seen on videotape buying alcohol for Beth at a bar close to campus, according to police. Maureen was originally to spend two days in jail. Eight months after Beth's death, Judge Robert Lyons amended charges from furnishing alcohol beverages to a minor to underage possession, a first-degree misdemeanor.

Maureen had to pay a $500 fine and participate in a countermeasure program (Gauthier, 2008; WKRC TV Cincinnati, 2008). However, her guilt of contributing to a friend's death would surely far outweigh her legal troubles.

DRIVING WHILE INTOXICATED

For nearly a century, alcohol has been recognized as one of the principal risk factors for motor vehicle crashes (National Institute of Drug Abuse, 1985). The percentage of alcohol-related motor vehicle fatalities did decrease from the early 1970s through the 1980s, partially thanks to the hard work of grass roots organizations such as Mothers Against Drunk Driving (MADD) to increase the penalties for this behavior and institute legal changes for limiting the amount of alcohol a person can have in their bloodstream and legally drive. However, that downward trend has now stabilized and may in fact be reversing (Wechsler et al., 2003). During the year 2002, over 17,000 motor vehicle fatalities in the United States involved alcohol, representing an average of one alcohol-related fatality every 30 minutes. The highest intoxication rates in fatal crashes were recorded for drivers 21-24 years old (National Highway Traffic and Safety Association, 2002). Most incidences of alcohol-impaired driving are never detected, with estimates that perhaps only 1% are reported (Flowers et al., 2008). Even so, approximately 1.4 million drivers were arrested in 2001 for driving under the influence, an arrest rate of 1 for every 137 drivers in the United States (NHTSA, 2002). College students are particularly prone to alcohol-impaired driving due to binge drinking practices (Flowers et al., 2008). The results of a recent national survey of self-reported behavior in the last 30 days indicated that 3 in 10 college students reported driving after drinking any amount of alcohol and 1 in 10 students reported driving after consuming five or more drinks (Wechsler et al., 2003).

Binge drinkers are 14 times more likely to drive while impaired by alcohol compared with non-binge drinkers (Naimi et al., 2003) and there is a strong association between binge drinking and alcohol-impaired driving (Flowers et al., 2008). The problem of impaired driving appears to even be more of a problem for binge drinkers than for chronic drinkers, as driving while intoxicated is more significantly associated with binge drinking than with chronic drinking (Borges and Hansen, 1993; Duncan, 1997; Everett et al., 1999; Flowers et al., 2008). As an example, an ecological analysis of the relationship between rates of binge drinking, chronic drinking, and driving while intoxicated (DWI) was examined using data from the Brief Risk Factor Surveillance System on 47 states. DWI rates were significantly associated with rates of binge drinking. DWI rates were not associated with chronic heavy drinking. Thus, if society has a goal to decrease DWIs, one approach might be to focus prevention efforts on reducing binge drinking (Duncan, 1997).

Despite the high rates of drinking and driving among college students, there is considerable variability in rates of alcohol-involved traffic fatalities across states. Wechsler et al. (2003) investigated whether policy factors are associated with alcohol-involved driving rates in college students. Using a national representative sample of students from colleges in 39 states, they administered questionnaires that examined driving after consumption of alcohol. They observed that the occurrence of drinking and driving among college students differs significantly according to the policy environment at the local and state level, as well as the enforcement of those policies. Lower rates of alcohol associated driving were observed with students attending college in states or localities that had laws restricting high volume sales of alcohol or laws targeting underage drinking, in combination with a strong investment in enforcement countermeasures. A similar study with the general U.S. population also observed that impaired driving rates coincide with state policies and their enforcement (Shults et al., 2002).

Strict policies against drinking and driving appear to reduce this risky and dangerous behavior in high school and college students (Hingson et al., 1994, 1996). For example, zero tolerance laws have been enacted in all 50 U.S. states. Wechsler et al. (2003) observed that underage college students had much lower rates of impaired driving than legal age students. It seems that the legal age students, who were less subject to strict drinking and driving laws, may be experiencing less perceived threat of consequences for driving after consuming alcohol. However, the authors of this study also observed that both underage and legal age college students were equally likely to ride with a drunk driver. Since riding with an intoxicated driver is not a violation of the law and doesn't expose oneself to legal risk, it appears that both demographic groups are likely to take this risk. It does suggest that having some sort of legal penalty for riding with an impaired driver by increasing the accountability of the passenger could potentially reduce impaired driving rates.

Reducing impaired driving and associated accidents might be achieved by further lowering the blood alcohol concentration (BAC) limits for driving. Wagenaar et al. (2001) questioned high school seniors in 30 states before and after BAC limits were lowered in 30 U.S. states between 1984 and 1998. They observed that the self-reported frequency of driving after any drinking declined 19% and the frequency of driving after drinking 5 or more drinks fell 23% with the change in law. However, the change in BAC limits only changed driving behavior. Despite decreases in impaired driving, the high school seniors reported no changes in overall amount of drinking or driving behavior in terms of overall miles driven. Fell and Voas (2006) reviewed 14 independent studies that measured changes in accident rates when the legal BAC limit was changed in various U.S. states. The authors observed that the lowering of the legal BAC limit from .10 to .08 resulted in 5-16% reductions in alcohol-related fatalities, injuries and crashes. All North American countries (Canada, U.S., and Mexico) and the United Kingdom currently have a .08 legal BAC limit. However, the legal limit is .05 BAC in numerous countries around the world (Australia, Austria, Belgium, Bulgaria, Croatia, Denmark, Finland, France, Germany, Greece, Israel, Italy, the Netherlands, Portugal, South Africa, Spain and Turkey). There is an international trend toward lower BAC limits. Lithuania has a limit of .04 and Norway, Poland, Russia, Sweden and Brazil have a limit of .02 BAC. Other countries, including the Czech Republic, the Dominican Republic, Hungary, Romania, and the Ukraine have effectively a zero limit (International Center for Alcohol Policies, 2008). Laboratory evidence suggests that critical driving functions begin to deteriorate at low BACs, such as .02 (Ogden and Moskowitz, 2004; Zador et al., 2000). The relative risk of being involved in a fatal crash as a driver is 4 to 10 times greater for drivers with BACs between .05 and .07 compared to sober drivers. Thus, lowering BAC limits may be in the best interest of the public (Desapriya, 2004; Fell and Voas, 2006). A variety of leading medical, crash prevention, public health, and traffic safety organizations from around the world support BAC limits at .05 or lower. These organizations include the World Medical Association, the American Medical Association, the British Medical Association, the European Commission, the European Transport Safety Council, the World Health

Organization, and the American College of Emergency Physicians (Chamberlain and Solomon, 2002).

Would the general public in North America support a lowering of BAC limits? Recent trends have moved toward increasing restriction on driving after drinking, with little public outcry. Canada adopted its current .08 BAC Criminal Code limit in 1969, with little objection. In the early 1970s, the first U.S. national efforts to counteract alcohol-impaired driving began. Laws were generally not based on a particular BAC but based on the presumption that the person was intoxicated, even though the police presumption of drunkenness could be refuted. The presumptive levels were high and generally set at .15 BAC. However, eventually most states adopted a .10 BAC limit. Since 1986, the Department of Transportation in the U.S. took the formal step of advocating a .08 BAC law by including it as one of the regulatory criteria for federal highway funding. U.S. Congress passed a law in 1995 requiring states to adopt zero tolerance laws for drivers younger than 21. Therefore, an individual who is under the age of 21, and not legally able to drink, will also be illegal if found driving with a BAC of .02 or .01 (depending on the state). Both enactment of lowered BAC limits and zero tolerance laws for underage drinkers have resulted in a reduced number of fatal crashes involving alcohol in the United States (Fell and Voas, 2006). Recent surveys in the United States indicate that most individuals believe they should not drive after two or three drinks (Royal, 2000). This is equivalent to a .05 BAC for many people (NHTSA, 1994). As such, the public may favor a BAC limit of .05, especially if this limit is explained to individuals in terms of consumption levels. In many countries that have already adopted a .05 BAC, this new limit has not resulted in extensive public outcry that the limit is too strict (Fell and Voas, 2006). However, countries such as Brazil recently lowered their limit to .02 with some public unhappiness about the change.

BINGE DRINKERS AND OTHER HEALTH RISKS

Adolescent binge drinkers take more risks when it comes to drinking and driving compared to their more moderate drinking and nondrinking peers. Not only are binge drinkers more likely to drive after drinking and ride with a drunk driver, but they are also more likely not to use seatbelts as a driver or as a passenger (Everett et al., 1999; Petridou et al., 1997). Adult heavy users of alcohol are also less likely to use seatbelts compared to the general public (Anderson et al., 1990; Reinfurt et al., 1996). In other words, binge drinkers compound their risk by not wearing seatbelts along with the fact that alcohol itself exacerbates the risk of serious injury and death in an automobile crash. Driving while impaired by alcohol or drugs and failing to use safety belts are the two most important risk factors for motor vehicle injury (US Preventive Services Task Force, 1996).

Binge drinkers are more likely to drive while impaired, ride with a drunk driver and not wear a seatbelt, all very risky behaviors, compared to their more moderate drinking peers. However, binge drinkers take numerous other risks and exhibit a syndrome of additional unhealthy behaviors. They are also more likely to use legal substances (such as cigarettes) and illegal substances (such as marijuana) (Wechsler et al., 1994, 1998, 2000). Binge drinkers are more likely to engage in unprotected and unplanned sex (Piombo and Piles, 1996;

Wechsler et al., 1994). Binge drinkers are more likely to get into trouble with campus police (Wechsler et al., 1994) or get hurt or injured (Wechsler et al., 1994; Wright and Slovis, 1996).

From birth until the middle of the fourth decade of life, the leading cause of death and disability is injuries. More than two-thirds of all deaths among young adults are a result of motor vehicle accidents, other unintentional injuries, homicides, and suicides. The most widely investigated risk factor for injuries is alcohol use. Field and O'Keefe (2004) examined the association of alcohol use, injury-related risk factors (such as driving or engaging in violence), and psychological characteristics with injury status. Data from 177 patients admitted to hospital for treatment of traumatic injury were compared with 195 general surgery patients as controls. They found that alcohol use is strongly associated with injury status. Once alcohol is taken into account, it appears that psychologically related variables (such as impulsivity, sensation seeking or risk perception) have limited utility in predicting injury status, similar to the findings of others (Cherpitel, 1993, 1996). Binge drinkers experience injuries because of their drinking and are also likely to be pushed, hit, or assaulted by a peer who has been drinking (Wechsler et al., 1994).

Binge drinkers take health risks in numerous domains, including their sexual behavior. Binge drinkers are likely to engage in unplanned sex (Wechsler et al., 1994). Binge drinkers also put themselves at increased risk for acquiring a sexually transmitted disease by not consistently using condoms (Kasenda et al., 1997). Students who are not binge drinkers but attend schools with high binge drinking rates reported more unwanted sexual advances compared to non-binge drinking students at schools with more moderated drinking (Wechsler et al., 1994).

The health risks associated with binge drinking are not limited to sexual behavior. Binge drinkers are more likely to be overweight or obese (Arif and Rohrer, 2005). Female adolescent and young adult binge drinkers are also more likely to engage in disordered eating (Pirkle and Richter, 2006). College student binge drinking has also been associated with a variety of other health-risk behaviors such as insufficient physical activity, infrequent breakfast consumption, fast food consumption and other behaviors that may lead to high stress levels, such as out of control spending and its associated high levels of credit card debt (Nelson et al., 2008). Thus, binge drinking may be one of a constellation of unhealthy lifestyle behaviors. Interestingly, this pattern has not been observed for moderate social drinking. For example, social drinkers (e.g., consumption of one or two drinks per day) are less likely to be overweight, compared to both non-drinkers and binge drinkers (Arif and Rohrer, 2005). Thus, while binge drinking may be associated with many risky health behaviors, the reverse may be true for moderate social drinking.

CONCLUSION

Binge drinkers engage in a variety of risky behaviors. Binge drinkers are more likely to drive after drinking, ride with an intoxicated driver and not wear a seatbelt. Lowered BAC limits and zero tolerance policies have reduced the number of alcohol-related automobile accidents and fatal injuries over the past few decades. However, this declining trend may now be reversing. Some evidence has suggested lowering the BAC limit for driving to .05 may be an effective means to counteract impaired driving in adults. A large proportion of impaired driving incidences involve binge drinkers. Therefore, programs that target both impaired driving and binge drinking should lower the rate of impaired driving. Binge drinkers also take health risks in other domains, such as behavioral risks that result in injuries, obesity and engaging in risky sexual behavior. If binge drinking were reduced, an improvement in healthy lifestyle might be achieved by many individuals.

Chapter 9

BINGE DRINKING AND EMERGENCY DEPARTMENT VISITS

For many adolescents and young adults, visits to the Emergency Department (ED) of a local hospital represent a routine consequence of their high-risk drinking habits. These presentations are often unflattering and potentially dangerous to both the patient and potentially their family as well. Alcohol-related visits to the ED are unfortunately all too common. Between the years 1992 and 2000, an estimated 68.6 million patients were seen in American EDs with an alcohol-related problem, at a rate of 28.7 per 1000 US population. Of even greater concern, during that eight year period the number of alcohol related visits increased by 18% (McDonald et al., 2004). This rise has also been seen elsewhere. Over a 4 year period in Ireland, patients presenting to an ED with blood alcohol concentration (BAC) > .08 g% rose 113%, while the number of patients with BAC > .48 g% rose 480% (Allely et al., 2006). This is not only concerning for the patients and families involved, but these alcohol-related visits place a significant burden on both human and material resources in the ED. Most ED presentations for alcohol-related issues occur after regular hours and on weekends, which coincides with peak demand for ED resources (Elder et al., 2004; Luke et al., 2002; Puljula et al., 2007). In adolescents, 82% of alcohol-related visits occurred after hours (Woolfenden et al., 2002). Not only are ED resources taxed, but these visits place a huge demand on prehospital care resources. Patients usually arrive in the ED by ambulance, particularly adolescents 10 to 18 years of age. 77% of intoxicated adolescents were transported to the ED by ambulance, a dramatically higher proportion than non-alcohol related visits (Woolfenden et al., 2002).

Although adolescents presenting to the ED with alcohol-related problems are fewer in number than their college-aged counterparts, the problem in this age group is far from insignificant. Estimated rates of ED visits among adolescents that involve alcohol varies from 3 to 5% (Colby et al., 2002; Elder et al., 2004; Sindelar et al., 2004), with a dramatic increase being seen in college student populations, where as many as 13-16% of ED visits are due to alcohol (Turner and Shu, 2004; Wright et al., 1998). In 2001, there were over 244,000 alcohol-related visits for patients between the ages of 13 and 25, with a staggering 49% of these involving patients under the legal drinking age of 21 (Elder et al., 2004). One of the most concerning findings is the number of visits among the youngest adolescents. In a review of ED visits to five Australian EDs by adolescents < 18 years of age, 23% were actually less than 14 years of age, with the youngest patient being 10 years old (Woolfenden et al., 2002).

This problem becomes even greater when other drug use is included. In a retrospective review of four different EDs in Australia, where 6% of ED visits among 12-19 year olds were secondary to alcohol use alone, this number increased to 15% when other drug use was included (Hulse et al., 2001). The important effect of the combination of alcohol and co-ingestions will be explored later in this chapter.

Among college-aged students, use of the ED for alcohol-related visits becomes significantly higher. The annual incidence of alcohol-related ED presentations is estimated to be anywhere from 1.7% to 3.9% of students per academic year. Once again, the younger members of this population seem to be at great risk for requiring visits to the ED. Among freshmen, the incidence is significantly higher at 2.9% to 9.3% of students per academic year (Wright et al., 1998; Wright and Clovis, 1996). As concerning as these numbers may seem, the reality of the problem becomes much more dramatic when these rates are adjusted over four years of college. At one typical US college, it was estimated that 1 out of 15 students will present to the ED with an alcohol-related disorder during their 4 years of college (Wright et al., 1998). This may at first glance seem to be specifically related to attendance at college. After all, over 50% of college students self-report getting drunk at least once a week during school (O'Brien et al., 2006) and the 48% of patients aged 18 to 29 years presenting to an ED who screened positive for alcohol problems were more likely to have college education (Odds Ratio 1.41), be a college student (OR 1.60), and cohabitate with friends (OR 1.19) (Horn et al., 2002). However, other studies have shown that problems related to alcohol use at this age may be an age-related phenomenon as opposed to a college phenomenon, in that patients 18 to 19 years of age who are not enrolled in college show similar patterns of alcohol consumption as their college-attending peers (Barnett et al., 2003).

REASONS FOR ALCOHOL-RELATED EMERGENCY DEPARTMENT VISITS

The majority of ED visits related to alcohol can be divided into two main sub-types: visits related to acute intoxication, and visits related to injury or trauma. Many ED visits related to alcohol use involve no external signs of injury, meaning the visits are purely related to drinking to excess only. The proportion of all alcohol related visits that are due to acute intoxication alone range from 33-38%, with no significant difference in these rates between adolescent and college aged patients (Barnett et al., 1998; Elder et al., 2004; Turner and Shu, 2004). In this intoxicated but uninjured sub-group, the patients tend to be younger, female and have higher alcohol concentrations than the injured sub-group (Allely et al., 2006; Colby et al., 2002). In one study of Canadian adolescents, the mean blood alcohol concentration of adolescents presenting to the ED was .20 g% with 90% exceeding the legal driving limit of .08 g% (Weinberg and Wyatt, 2006).

These visits for severe intoxication can be associated with significant morbidity or medical problems directly arising from their intoxication. Many of these patients present with a significantly depressed level of consciousness, some of which are so depressed that they no longer have an intact gag or cough reflex. Without these basic reflexes, and given the predisposition to vomiting when acutely intoxicated, these patients are at a significant risk of choking or aspirating stomach contents into their lungs. In order to prevent this, they may require endotracheal intubation, which involves insertion of a tube into their airway, in order to protect their lungs. Of the Canadian cohort listed above, 24% had a level of consciousness where this treatment decision would be recommended (Weinberg and Wyatt, 2006).

The medical consequences of acute intoxication are not limited to risk of vomiting and aspiration. When young patients are brought to the ED with a severely depressed level of consciousness, even if they are known to have been drinking alcohol, or have a BAC that would explain this, it is often necessary to ensure that there are no other reasons for their presentation. This is even more true given the association between intoxication and injury, as discussed later. Severely intoxicated and unconscious patients cannot provide reliable information to physicians, and their physical exam findings can be significantly affected by their intoxication.

Therefore, they may be subjected to unnecessary tests in order to ensure they haven't sustained a significant injury or ingested other substances. As a typical example of this, patients may require a CT (computed tomography) scan of their head in order to ensure they haven't had a fall with a resulting significant head injury. Tests such as these are not without their side effects or complications. Unnecessary CT scans expose patients to radiation, which may potentially be associated with an increased lifetime risk of cancer. Beyond the medical consequences of such test, there are often significant costs associated with these investigations and hospitalization. As many as 1 in 20 intoxicated adolescents will require admission to hospital (Turner and Shu, 2004).

The other sub-group of alcohol-related ED visits is for injury or trauma. Among adolescents and young adults who present to the ED with injury, alcohol is involved in anywhere from 32% to 50% of cases (Colby et al., 2002; Loiselle et al., 1993; Porter, 2000; Rivara et al., 1992; Sindelar et al., 2004). The prevalence of alcohol among patients with injury or trauma increases steadily through the adolescent to college age groups, with rates increasing from 13% among adolescents to 47% among college aged young adults (Porter, 2000).

Although this would suggest that the problem of alcohol related injury is greater among college aged patients, other studies have found younger patients to be as susceptible. The mean age of patients presenting with alcohol related injury has been reported as low as 15 years (Loiselle et al., 1993; Meropol et al., 1995). Despite the fact that these patients are presenting for their injuries, not specifically for their intoxication, they are still at risk for all of the medical problems described above. As many as 64% of patients presenting with alcohol-related injuries have BACs above .10 g% (Porter, 2000; Rivara et al., 1992). The association between alcohol and injury is not only limited to the time of acute intoxication. Young trauma patients, particularly male patients, are significantly more likely to screen positive for hazardous alcohol consumption, regardless of whether their current injury seems to be immediately due to alcohol consumption (Tjipto et al., 2006). It appears that dangerous drinking behavior, regardless of timing, may be associated with an increased risk of injury or trauma.

With alcohol-related injury and trauma comprising the majority of alcohol related ED visits, and with the high resource burden of these injuries on the ED, a significant amount of research has been done in this area. The rest of this chapter will explore what is understood about the association between alcohol and injury requiring medical intervention.

GENERAL INJURY RISKS RELATED TO ALCOHOL USE AND ABUSE

There is ample evidence suggesting an association between alcohol use and the risk of injury. A recent review of the literature revealed that adolescents that were positive on screening for alcohol were more likely to be injured, to have a prior history of injury, to require trauma service care and to have injury complications, when compared to alcohol-negative peers (Sindelar et al., 2004). In a recent review of 8,000 patients from 28 studies representing 16 countries, 24% of patients with a positive blood alcohol screen and 29% of patients who self-reported alcohol use were found to have an alcohol-related injury. The review also found rates of alcohol-related injury were highest in groups with higher overall average consumption, higher rates of detrimental drinking patterns, and in populations where the legal blood alcohol level while driving was higher (Cherpitel et al., 2005). The risk of injury is significantly greater when alcohol has been consumed recently. As many as 35% of injured patients seen in some EDs have consumed alcohol prior to sustaining their injury, which is almost 3 times higher than in uninjured patients (Humphrey et al., 2003; Weinberg and Wyatt, 2006).

These trends are seen consistently among college students. Up to half of the college-aged students seen in an ED while intoxicated have been using alcohol, with the pattern of their drinking playing a role in how likely they are to require emergency care for an injury (Colby et al., 2002; Wright and Clovis, 1996). Students who reported getting drunk at least once per week were 5 times more likely to be injured, compared with students who drank less. These students were also more likely to cause an injury requiring medical treatment to someone else, including being twice as likely to cause an automobile crash, a burn, or a fall (O'Brien et al., 2006).

Some have suggested, however, that the relationship between traumatic injuries and alcohol use is overstated and that traumatic injuries are more common in youth in general, based on high risk behavior, feelings of invulnerability, and lack of experience with driving, among others. Although there is no doubt that injuries are indeed more common among youth, this argument has been dispelled by a large multi-national study of over 18,000 patients from EDs in seven countries (Argentina, Canada, Italy, Mexico, Poland, Spain, United States). In this study, after controlling for gender and age, blood alcohol concentration was found to be highly predictive of an injury, with a positive blood alcohol screen being associated with a 1.5 times greater risk of sustaining an injury (Cherpitel et al., 2004).

There also appears to be a significant link between the patient's sex and the risk of alcohol-related injury. In general, the risk of injury for women is significantly elevated for any consumption of alcohol. For men, the risk seems to be increased only when alcohol consumption exceeds 90 grams, which is equivalent to approximately 6 cans of beer (Stockwell et al., 2002). In college students, injury and trauma occurred more frequently among men greater than 18 years old (Puljula et al., 2007).

The types of injuries seen among intoxicated patients are highly variable. Obviously motor vehicle collisions are of significant concern because of the much higher likelihood of grave and life-threatening injuries at the hands of an intoxicated driver. However, some of the injuries seen are not necessarily sustained as part of the typical crashes related to drunk drivers. Passengers who are found to be intoxicated seem to be at just as high a risk of needing future emergency care for an alcohol related injury as drivers who are intoxicated, and passengers have even been found to be at similar risk of death in the future due to intoxication-related injury (Schermer et al., 2001). Although the reasons for this are not clear, it may be that the type of risk-taking behavior seen among drivers who are willing to get behind the wheel while intoxicated are also seen in passengers who socialize with these drivers. These passengers may also be as more likely to drive themselves while intoxicated, or to partake in other risky behaviors after consuming alcohol.

The risk of significant injury following alcohol consumption is also seen in other modes of transportation, including cycling. While the risk to others is less than in motor vehicle collisions, the risk to the cyclist in particular is very high, considering the fact that cyclists tend to share the road with motor vehicles, and that they are highly exposed without the protective covering of a car if they are involved in a collision. Furthermore, intoxicated cyclists are far less likely to wear a helmet compared to cyclists who are not intoxicated, with rates decreasing from 35% to 5% after alcohol use (Li et al., 2001). Among cyclists presenting to a Baltimore area ED with a cycling related injury, 13% were found to have a positive blood alcohol concentration. The severity of injuries among intoxicated cyclists is also significantly higher than is seen in injured cyclists who have not been drinking. In this same study, 23.5% of intoxicated cyclists presenting to the ED were fatally injured and a further 9% were seriously injured, significantly higher than the 3% of cyclists with no alcohol consumption. Cyclists who had a BAC of .02 g% were 5 times more likely to sustain an injury than cyclists who had not consumed alcohol, with this risk increasing to 20 times when BAC exceeded .08 g%, the legal limit of alcohol for driving (Li et al., 2001).

SPECIFIC INJURY RISKS RELATED TO FEATURES OF ALCOHOL CONSUMPTION

There are specific alcohol-consumption factors that further predispose intoxicated patients to injury. The most significant appears to be binge drinking related to both the timing and the volume of alcohol ingestion. With respect to the timing of alcohol ingestion relative to an injury, studies have found that consuming any alcohol in the previous six hours significantly increases the risk of injury, in some cases over two times (Bazargan-Hejazi et al., 2007; Watt et al., 2004). The risk seems to increase the closer the consumption of alcohol occurs to the injury, with the injury risk increasing by over 4 times if alcohol has been consumed in the previous hour (Borges et al., 2004). There also appears to be a link between the volume of alcohol ingested and the risk of injury. In general, studies with higher overall rates of consumption report a higher rate of alcohol-related injury (Cherpitel et al., 2005).

These two factors are certainly linked. The greater the number of drinks consumed closer to the time of the injury, the greater the risk of injury. When only one or two drinks were consumed in the previous six hours, the risk of injury increased by about 1.5 times. However, when 3 or 4 drinks were consumed, this risk increased to 4 to 6 times greater than baseline, and up to 13 times greater after 6 drinks. All of this suggests that binge drinking patterns are much more likely to be associated with injury than usual social drinking. The more likely a consumption pattern is to result in a higher BAC, that is a larger number of drinks over a shorter period of time, the more likely the drinker is to sustain an alcohol related injury. In keeping with this observation, patients with chronic alcohol abuse have been found to be less likely to be injured, further adding to the evidence that acute binge drinking is associated with injury (Vinson et al., 2003).

These effects seem to be independent of sex and ethnicity. After controlling for differences in acute intake based on sex (> 40 grams in females, > 60 grams in males), Australian researchers found the risk of injury was still increased by a factor of 2.5 (Watt et al., 2004). The relationship between binge drinking and injury seems to be consistent among populations around the world, with similar results being seen in studies from many different countries. In Sweden, the proportion of people who reported heavy episodic drinking (> 6 glasses or 72 g) once a month or more was nearly twice as high among injured patients as in the general population (Nilson et al., 2007). Another Swedish study reported that risky drinkers (defined as weekly consumption > 80 g (women) and > 110 g (men) and/or heavy episodic drinking (> 6 or more glasses = 12 g) on one occasion at least once a month) were six times more likely (OR 6.4) to be injured, independent of the activity being performed (Nordqvist et al., 2006). In Switzerland, binge drinking (≥5 drinks for men and ≥4 drinks for women at least once per month) was closely associated with current injury in a dose-response relationship (Gmel et al., 2007). This increased risk was particularly seen in patients who were otherwise low volume drinkers experiencing heavy episodic drinking as described above (Gmel et al., 2006). A study performed in New Zealand found that in injured patients who admitted to alcohol use pre-injury, the median amount of self-reported alcohol consumed was 103 mL, which they calculated as being equivalent to the ingestion of approximately seven cans of beer. Based on these results, they were able to calculate a cumulative injury risk of 1.14 for each 30 mL of absolute alcohol (approximately two cans of beer) consumed (Humphrey et al., 2003).

The type of alcohol may also play a significant role in the risk of injury. An Australian study found that the risk of injury was three times higher with the ingestion of spirits or a combination of spirits and other types of alcohol, compared with a two times higher risk of injury with the ingestion of beer alone (Watt et al., 2004). This has particular significance among adolescents and young adults, where spirits have been shown to be the alcohol of choice in this age group (Weinberg and Wyatt, 2006).

The association of binge drinking with injury is particularly noticeable in groups who are not normally heavy drinkers. In the large multi-national study mentioned previously, the overall likelihood of injury was found to be far less for heavier regular drinkers than for lighter drinkers who binged less frequently (Cherpitel et al., 2004). This has also been seen in other studies where patients with alcohol dependence and with high frequency of drunkenness had lower risks of injury than patients with the lowest usual involvement with alcohol and with lower self-reported episodes of drunkenness in the last year (Borges et al., 2004). There are several possible reasons for this finding. One may be that heavy regular drinkers have greater tolerance, particularly in their ability to adapt their behavior, than binge drinkers. Regular drinkers may be less likely to take unnecessary risks and may therefore avoid situations that are likely to lead to injury. The second reason may be related to the typical populations where these drinking behaviors are seen. Binge drinking is much more common among adolescents and college aged students, which also tends to be the age range where perception of risk is lowest and risk taking behavior more likely.

Once again, sex differences may also play a role. Low-volume male drinkers with heavy episodic drinking had more alcohol-related injuries than low-volume female drinkers with heavy episodic drinking (46.9% vs. 23.2%) (Gmel et al., 2006). Whether this is due to differences in likelihood of engaging in risky behaviors between males and females alone, or some other factor, is not clear. However, the impact of this pattern of drinking on risk of injury overall is of great concern for adolescents and young adults, who are relatively inexperienced with alcohol ingestion and whose pattern of drinking is much more likely to reflect infrequent binge drinking than heavier regular drinking.

As mentioned earlier, the risk of injury may be impacted by the presence of co-ingestants. In a study of adolescents aged 12 to 19 years, 15% of presentations to an ED were due to alcohol or other drugs. Alcohol was the most common ingestant, accounting for 41% of these presentations, followed by heroin (15%) and prescription drugs and over the counter preparations (15%). Over 30% of these adolescents were brought to the ED because of an injury, and this was significantly more likely following alcohol use than other substances (Hulse et al., 2001). Although not classically considered a co-ingestant, the impact of combining caffeine and alcohol may be significant. Consumption of alcohol mixed with caffeinated energy drinks is popular on college campuses. Limited research suggests that energy drink consumption lessens subjective intoxication in persons who also consumed alcohol. Caffeinated drinks also allow drinkers to remain alert for longer and ingest greater amounts of alcohol. In college students from 10 universities in North Carolina, 24% reported consuming mixed energy drinks with alcohol in the past month. Mixed energy drinks and alcohol were associated with significantly more days of heavy episodic drinking per month (6.4 days/month vs. 3.4 days/month) and twice as many episodes of weekly drunkenness (1.4 days/week vs. 0.73 days/week). Students who reported consuming energy drinks and alcohol also reported significantly higher prevalence of alcohol-related consequences including being physically hurt or injured, and requiring medical treatment, or participating in activities known to increase the risk of injury, such as being a passenger in a car driven by an intoxicated driver (O'Brien et al., 2008; see Chapter 16 for more details).

Finally, the volume of alcohol consumed while binging may also be related to the type and severity of injury that is sustained. Patients with a blood alcohol concentration .20 g% were most likely to have had minor soft tissue limb injuries (58%), while patients with a blood alcohol concentration between .20-.25 g% were most likely to have had significant limb fractures (55%), and patients with a blood alcohol concentration > .25 g% were most likely to have had significant head injuries (90%) (Johnston and McGovern, 2004). Specific injuries related to alcohol use will be discussed in the following section.

SPECIFIC ALCOHOL-RELATED INJURIES: A REVIEW

Alcohol consumption may not only impact the likelihood of injury, it may actually impact the type and severity of injuries as well. Several studies have looked at this issue. Among trauma patients, a numerical score, called the Injury Severity Score (ISS), is calculated in order to quantify not only severity of injuries, but also extent of injuries. A higher score indicates that injuries are more severe and involve a greater number of body systems. One study which investigated the association between alcohol consumption and injury found that not only was alcohol associated with a 1.5 times greater likelihood of injury, but that the severity of injury as measured by the ISS was 30% higher among patients who had consumed alcohol prior to their injury compared to patients who had not consumed alcohol. This increase in ISS was found to be independent of increasing blood alcohol concentration, meaning that the presence of alcohol alone may place patients at a more significant risk of severe injuries (Waller et al., 2003). This relationship between alcohol intoxication and severe injury was also found in studies looking specifically at motor vehicle crashes, where intoxicated drivers were found to have not only a significantly higher overall ISS, but also a lower Glasgow Come Scale score (which measures a patient's level of consciousness, with higher scores associated with an increased level of alertness) and a poorer systemic blood pressure. Of all the variables that were investigated for their possible relationship with injury severity, alcohol intoxication was the only one that independently predicted a higher risk of medically significant consequences (Fabbri et al., 2002; Shih et al., 2003).

The relationship between consumption of alcohol and specific type of injury is less clear. Several studies report no overall association between alcohol consumption and the distribution of specific body regions injured (Fabbri et al., 2002; Watt et al., 2005), while others have noted more prevalent injury patterns, particularly head injuries, after alcohol consumption. One study from Northern Ireland found that intoxicated trauma patients had a significantly higher incidence of head injury than trauma patients with no evidence of alcohol intake (48% vs. 9%) (Johnston and McGovern, 2004). This was similarly seen in a review of Canadian adolescents presenting to the ED while intoxicated, where head injuries were the most common injury seen at 42% (Weinberg and Wyatt, 2006). There may also be an association between alcohol-induced trauma and the severity of head injuries. A review of trauma patients in Michigan revealed that patients with a positive blood alcohol level were twice as likely to have a severe traumatic brain injury, diagnosed using a CT scan of the brain, as patients who had not tested positive for alcohol, even after controlling for crash severity (Cunningham et al., 2002). Other neurological injuries, such as spinal cord injury, have also been linked with alcohol intake (Forchheimer et al., 2005).

Another area where an association with alcohol intoxication has been found is intentional injuries, or injuries related to assault. Alcohol consumption has been found to be more strongly associated with ED visits related to assault and intentional injury than to visits for unintentional injuries or injuries of unknown intent (Elder et al., 2004). Intentional injuries make up a significant proportion of alcohol related injuries, with as many as 16% of ED visits for alcohol related injuries falling in this category (Colby et al., 2002; Humphrey et al., 2003). This relationship has also been seen in studies specifically related to adolescents (Meropol et al., 1995), as well as studies related to nightclub attendance (Luke et al., 2002), which is more relevant to college-aged students. The types of injuries caused by assault and intentional violence can differ significantly from unintentional injuries. Lacerations and bite wounds are much more common, as are wounds related to striking another person's mouth and teeth with one's fist. Lacerations were the most common injury found from patients presenting following attendance of a nightclub, with the face being the most common area of the body affected (Luke et al., 2002). The association between alcohol consumption and bite wounds is particularly striking, with alcohol being involved in up to 86% of bites. This is of particular significance for medical treatment, as bite wounds have a high morbidity due to a significant infection rate. In order to prevent infection, wounds often must be left open rather than stitched closed, in order to prevent closing in difficult-to-treat bacteria that originate in the human mouth. These wounds take longer to heal, are more likely to result in significant scars, and are much more likely to become further infected with more common bacteria as they heal (Henry et al., 2007).

Alcohol consumption is not only a linked with intentional injuries to others, but also with deliberate self-harm. There is a higher frequency of alcohol use among patients seen in the ED with self-inflicted injuries than among patients seen for unintentional injuries or injuries of unknown intent (Elder et al., 2004). Significantly higher numbers of positive toxicology screens, which identify other ingestants or co-ingestants with alcohol, have also been seen with adolescents presenting to the ED with an intentional injury (Li et al., 1999; Loiselle et al., 1993). The link between self-harm and alcohol use is very strong, with consumption of alcohol preceding nearly 50% of these injuries (Li, 2007).

Finally, although unrelated to traumatic injury per se, there appears to also be a link between alcohol intoxication and the risk of cardiac injury or abnormal heart rhythms. Electrocardiograms (ECGs) are used to assess heart rhythm as well as to look for signs of injury to the heart muscle, such as that seen during a heart attack or severe angina. Acutely intoxicated patients have been found to have abnormalities on their ECGs which may predispose them to dangerous cardiac arrhythmias (Aasebo et al., 2007). Although these changes would only be expected to be seen in adult patients with a predisposition to heart disease, this has not been the case. Previously healthy patients as young as 19 years old have been found to have ECG findings of a heart attack after drinking alcohol (Biyik and Ergene, 2006). Although much less common than vomiting and choking, it is possible that toxic effects of alcohol on the heart may be the cause of sudden death in adolescents and college-aged students after binge drinking.

LONG TERM OUTCOMES OF ALCOHOL-RELATED ED VISITS

Alcohol-related visits to the ED are clearly not uncommon among adolescents and college aged students, and often are related to the significantly increased risk of injury while intoxicated. But is there a link between an ED visit in the present and future risk of alcohol-related problems? Can we use this ED visit as a way to begin interventions aimed at reducing the long-term problems of recurrent or chronic alcohol abuse? Some studies have looked at these issues in order to determine whether there is in fact a link between a current ED visit and long-term outcomes. The results are worrisome. There are indeed higher rates of chronic problem drinking among older adolescents who were previously treated for acute alcohol intoxication. While only one visit to the ED with a positive alcohol blood test certainly predicts future problems, the likelihood of more severe drinking problems in the future increases with increasing frequency of drunkenness as well as with being treated for assault (Barnett et al., 2003; Kelly et al., 2004). However, this predisposition to problem drinking does not appear to be simply a short-term phenomenon linked to being young and foolish. Very few adolescents between 13 and 19 who have an ED presentation for alcohol present again within 6 months, unless they are polysubstance users (Tait et al., 2002). Perhaps this initial visit to the ED scared younger adolescents into avoiding alcohol for a period of time, but without the protective effect enduring. Unfortunately, it seems that adolescents who present with higher alcohol levels at first presentation to the ED eventually will be seen again for alcohol-related issues (Baune et al., 2005), and heavy binge drinkers are even more at risk, with the binge pattern of drinking being associated with multiple ED visits over time (Cherpitel et al., 2006). While it may seem that this first presentation is an opportunity to intervene, not all patients presenting to the ED are tested or asked about their alcohol consumption related to their visit. In a 5-year review of the American National Trauma Data Bank, only about one half of all trauma patients are tested for alcohol, although one half of those alcohol tests are positive (London and Battistella, 2007). In adolescent patients aged 13 to 19 years presenting to an ED, testing was performed more commonly if the patients were male, older, injured, treated during overnight shift or on weekends, involved in a motor vehicle crash or assault, or appeared clinically intoxicated (Puljula et al., 2007). This bias in testing raises the question of whether at-risk patients who may benefit significantly from intervention are being missed. Routine screening for alcohol during ED visits has been suggested for both adolescent and injury presentations to the ED. Although the benefits and costs of such a suggestion have yet to be worked out, the ED visit may present a real opportunity for screening and intervention in order to prevent long-term consequences of alcohol use and abuse.

CONCLUSION

Visits to the Emergency Department of a local hospital are often routine consequences of binge drinking in adolescents and college students. Alcohol-related visits have increased in recent years, placing considerable burden on the health care system. The majority of visits are either directly related to acute alcohol intoxication or for treatment of an injury or trauma sustained during drinking. Alcohol consumption impacts the likelihood of injury, as well as the type and severity of injury. An Emergency Department visit can be an important risk factor for future problems with alcohol. There are higher rates of chronic problem drinking among adolescents who were previously treated for acute alcohol intoxication. Furthermore, heavy binge drinkers are even more at risk, with this pattern of drinking being associated with multiple visits over time. Thus, any presentation in the Emergency Department is an important opportunity to intervene in reducing hazardous drinking behavior, though currently only half of all trauma patients are tested for alcohol. The Emergency Department may present an important location for screening and intervention for high school and college drinkers to prevent severe long-term consequences of alcohol use and abuse.

Chapter 10

SCREENING AND BRIEF INTERVENTIONS IN THE EMERGENCY DEPARTMENT

The 500,000 college aged students who are injured every year while under the influence of alcohol could be viewed as failures of the primary prevention strategies currently being used in schools and on college campuses. Perhaps these students somehow escaped all exposure to these programs. More likely, they didn't feel the message of these programs was relevant to them, or that their drinking was really as risky as these programs try to portray. Regardless of the reason why these strategies haven't worked for these students, is there still a chance to "get through" to them before they are permanently injured or even become one of the 1,400 who die each year from alcohol-related injuries?

Adolescents and young adults tend to be some of the greatest risk-takers, not just with alcohol. Often, youth believe that risks are either overstated, or simply highly unlikely to happen to them. If the problem is simply lack of acceptance of the truth of these risks, the experience of an adverse event related to their drinking may act as a powerful means to change this attitude. Among 18-30 year olds presenting to the emergency department with alcohol-related problems, 76% had not previously been concerned about their alcohol use and had never been advised to cut down (Tjipto et al., 2006). College students interviewed after requiring transport by ambulance to the hospital for an alcohol-related issue said they didn't previously see themselves as at risk but now recognized the risk and planned to change their behavior (Reis et al., 2004).

Does the fact that a youth needs medical care for an alcohol-related problem make them more receptive to intervention and change? Is there a way to capitalize on this opportunity while they are in the emergency department, and intervene to help facilitate changing their drinking behavior and avoid future emergency department visits, injuries or even death?

SCREENING

In order to maximize the potential benefit from an emergency department based intervention, there needs to be a process to identify all possible patients that will benefit. Identifying the 'at risk' drinker is no easy task. Blood tests for alcohol levels or signs of alcohol-related organ damage, breathalyzers, and saliva tests have all been used to identify acutely intoxicated patients, or to look for evidence of organ damage related to excessive alcohol use. The problem, especially for adolescents and college aged students, is that their pattern of drinking may not necessarily yet have led to signs of organ damage, and these other methods of screening will only identify those who are acutely intoxicated when they present to the emergency department. In addition, it is not practical to test with screening lab work every patient who presents to the emergency department, which leaves the decision on who to test up to the emergency physician. Clinical suspicion, unfortunately, has been shown to have very poor sensitivity and specificity (Gentilello et al., 1999). In one study, only 50% of trauma patients were tested for alcohol use despite much higher known rates of alcohol involvement in traumatic injuries (London and Battistella, 2007). In another study of 13-19 year olds, emergency department physicians were more likely to test for alcohol if the patient was older, male, injured or presenting overnight or on weekends. However, in this age group, young female patients were more likely to be treated for intoxication alone (without an injury) and tended to present more evenly across all shifts (Colby et al., 2002). Screening in this way is clearly problematic.

In order to identify all patients whose drinking places them at risk, structured questionnaires may be much more useful than laboratory testing. Multiple questionnaires have been tested for their ease of use as well as their sensitivity and specificity for alcohol use disorders. However, a significant proportion of alcohol use disorders in the adult population are characterized by alcohol dependence rather than the hazardous or at-risk patterns of drinking more common in adolescents and college-aged patients. The most effective approach for identifying alcohol dependence is not necessarily the most effective for identifying the hazardous drinking behaviors seen in adolescents and young adults, such as binge drinking, which makes extrapolation of the results of studies in the adult population questionable. For example, the CAGE questionnaire (Ewing, 1984) was found to be 76% sensitive and 90% specific for alcohol dependence (Bernadt, 1982). This questionnaire consists of four questions: Have you ever felt you should Cut down on your drinking? Have people Annoyed you by criticizing your drinking? Do you ever feel bad or Guilty about your drinking? and Have you ever had a drink first thing in the morning (an Eye-opener) to steady your nerves or get rid of a hangover? Despite the ease of administering this screening questionnaire and its success in identifying alcohol dependence, it has fared far less well compared to other screening questionnaires for identifying at-risk and hazardous drinking behavior more typical in youth (Chung et al., 2000; Kelly et al., 2004). A somewhat longer questionnaire, the Alcohol Use Disorders Identification Test (Saunders et al., 1993), was developed to identify at-risk and hazardous drinkers in addition to the dependent drinker. Shown in Table 10-1, it consists of 10 questions which need to be scored, and various cut scores have been used as positive screens. It has been shown to work very well in particular for the adolescent population (Chung et al., 2000; Kelly et al., 2004), but its length and need for scoring make it somewhat less ideal for the emergency department setting.

New combined approaches have been explored to achieve the effectiveness of longer screening questionnaires such as AUDIT in more brief, usable formats for the emergency department (D'Onofrio et al., 2006; Kelly et al., 2004). For example, the American College of Emergency Physicians has adopted a combined approach using 3 questions regarding quantity and frequency of drinking proposed by the National Institute of Alcohol Abuse and Alcoholism (NIAAA, 2006).

Table 10.1. The AUDIT Questionnaire

1. How often do you have a drink containing alcohol?
    - Never
    - Monthly or less
    - 2–4 times a month
    - 2–3 times a week
    - 4 or more times a week

2. How many standard drinks containing alcohol do you have on a typical day when drinking?
    - 1 or 2
    - 3 or 4
    - 5 or 6
    - 7 to 9
    - 10 or more

3. How often do you have six or more drinks on one occasion?
    - Never
    - Less than monthly
    - Monthly
    - Weekly
    - Daily or almost daily

4. During the past year, how often have you found that you were not able to stop drinking once you had started?
    - Never
    - Less than monthly
    - Monthly
    - Weekly
    - Daily or almost daily

5. During the past year, how often have you failed to do what was normally expected of you because of drinking?
    - Never
    - Less than monthly
    - Monthly
    - Weekly
    - Daily or almost daily

6. During the past year, how often have you needed a drink in the morning to get yourself going after a heavy drinking session?
    - Never
    - Less than monthly
    - Monthly
    - Weekly
    - Daily or almost daily

Table 10.1. (Continued)

7. During the past year, how often have you had a feeling of guilt or remorse after drinking?
    - Never
    - Less than monthly
    - Monthly
    - Weekly
    - Daily or almost daily

8. During the past year, have you been unable to remember what happened the night before because you had been drinking?
    - Never
    - Less than monthly
    - Monthly
    - Weekly
    - Daily or almost daily

9. Have you or someone else been injured as a result of your drinking?
    - No
    - Yes, but not in the past year
    - Yes, during the past year

10. Has a relative or friend, doctor or other health worker been concerned about your drinking or suggested you cut down?
    - No
    - Yes, but not in the past year
    - Yes, during the past year

Scoring:
Scores for each question range from 0 to 4, with the first response for each question (e.g. never) scoring 0, the second (e.g. less than monthly) scoring 1, the third (e.g. monthly) scoring 2, the fourth (e.g. weekly) scoring 3, and the last response (e.g. daily or almost daily) scoring 4. For questions 9 and 10, which only have three responses, the scoring is 0, 2 and 4 (from left to right).
A score of 8 or more is associated with harmful or hazardous drinking.
A score of 13 or more in women, and 15 or more in men, is likely to indicate alcohol dependence.

Once establishing that the patient drinks any alcohol at all, the following three questions are asked: On average, how many days per week do you drink alcohol? On a typical day when you drink, how many drinks to you have? and What is the maximum number of drinks you had on a given occasion in the last month? If any of the answers fall above low risk guidelines (<14 drinks per week or 4 drinks per occasion for men, and <7 drinks per week and 3 drinks per occasion for women), the CAGE questionnaire is then administered. Despite the relative ease of administering such combined approaches, few studies have used these methods for screening and outcomes from interventions based on these methods are not yet clear.

SCREEN POSITIVE....NOW WHAT?

The time and effort involved in administering screening tests and questionnaires is only worthwhile if a successful intervention is available once at-risk drinkers have been identified. Options for intervention vary from referral to social services or drug and alcohol services to brief targeted intervention right in the emergency department. Studies looking at referral patterns and compliance with follow-up have shown disappointing rates of both, thus raising several questions. First, does referral for intervention after the emergency department visit actually happen consistently enough to be an effective strategy? And, second, does the motivation for the patient to attend wane as the effects of the emergency department visit start to fade from memory?

These questions have led to the strategy of intervening right in the emergency department itself. With the time constraints and challenges of a busy emergency department, an intervention that occurs here must be relatively brief to be realistic, while also capitalizing on the potential motivating effects of being in the emergency department for an alcohol-related reason. Interventions that have been tried include simple verbal advice-style feedback, generic or specific written material in the form of pamphlets of written discharge instructions regarding drinking, or more structured interview-style interactions. This latter approach, variably referred to as brief intervention, brief negotiation interview, or brief motivational interview, uses the emergency department visit as an opportunity to provide direct, one on one counseling at a potentially highly influential time. This brief intervention approach has shown the greatest promise, and is now the focus of increasing research in secondary prevention of alcohol-related illness and injury.

A brief intervention is an interaction between the patient and a physician, nurse, or designated alcohol intervention worker such as a psychologist or social worker. The interaction can range from a few minutes to up to an hour, and usually incorporates feedback, advice, and techniques to enhance motivation to change drinking behavior in the patient. The goal of the interaction is to guide the patient towards reducing their drinking behaviors to within low-risk guidelines in order to decrease the risk of future illness, injury, or emergency department visits (D'Onofrio et al., 2006). This approach was first adapted for the emergency department in 1994 at Boston Medical Center (Bernstein et al., 1996) and has been studied in various forms and in multiple populations extensively since then. The four basic steps of the interaction have been described by D'Onofrio and colleagues (D'Onofrio et al., 2005). First, the subject of the patient's alcohol consumption is raised, and permission is requested to further discuss this drinking. Second, feedback is provided on how the patient's drinking levels compare with recommended low risk levels, and the effects of drinking above this level are discussed. The patient is asked about their perception of the connection between his or her drinking and this emergency visit, and if the patient does not make the connection this information can be provided. Third, the patient's motivation to reduce drinking is explored. The practitioner asks the patient to expand on these reasons in order to reinforce and enhance the motivation to change. Finally, a plan of action is negotiated. Realistic goals are set with the patient, and these can then be documented on a discharge instruction sheet or other written agreement.

Not surprisingly, with the varying target populations and the variable styles of delivery, the results of studies of the effectiveness of brief interventions have been mixed. Multiple studies have shown that any types of intervention, not just the brief intervention as described, leads to a decrease in alcohol consumption (Daeppen et al., 2007; Dent et al., 2008; D'Onofrio et al., 2008; Sommers et al., 2006). There are several possible reasons for this observation. The need to come to the emergency department alone may be a powerful motivator for change regardless of how this is addressed in the emergency department. In addition, the questions asked in the screening process may lead to greater insight on the part of the patient to the deleterious effects of his or her current drinking behavior, thus providing enhanced motivation to change. However, the additional benefit of brief intervention over and above these inherent motivators has been shown in some higher-powered studies. Brief intervention has been shown in a number of studies to have a greater effect than other interventions or screening alone not only in reducing alcohol consumption (Academic ED SBIRT Research Collaborative, 2007; Gentilello et al., 1999), but also in decreasing drinking and driving, alcohol-related injuries, and emergency department presentations for alcohol-related problems over the 6 to 12 month period following the intervention (Baird et al., 2007; Gentilello et al., 1999).

But what about adolescents and college-aged students? Do they benefit from such emergency department based interventions? Few studies have looked at the adolescent population specifically. However, the small number of studies that have been done are encouraging. Brief intervention decreased future presentations to the emergency department for alcohol-related issues and led to increased attendance at treatment programs (Tait et al., 2005). In another study, motivational interviewing led to a decrease in number of drinking days per month and a decrease in binge drinking over the following 12 months among adolescents who screened positive for more problematic alcohol use at baseline (Spirito et al., 2004). In college aged students, motivational interviewing has not only led to a decrease in overall alcohol consumption and binge drinking behavior (Monti et al., 1999, 2007), it has also been effective in decreasing rates of drunk driving and alcohol-related injuries (Monti et al., 1999). The receptiveness of college students to emergency department based intervention was found to be very high, with 96% of screen positive students accepting counseling during their emergency department visit, and at least 75% reporting that they had found the counseling to be helpful (Helmkamp et al., 2003).

Given the variability of benefit of brief intervention over standard care, is it possible to identify who will benefit and who is less likely to benefit from brief intervention? Motivation to change has been shown to be the greatest predictor of success in both college-aged patients (Barnett et al., 2006; Leontieva et al., 2005) and adolescents (Barnett et al., 2002). Experiencing negative consequences of alcohol such as injury (Reed et al., 2005) or need for transport to hospital (Reis et al., 2004) has been shown to lead to greater readiness for change, which explains why simply coming to the emergency department leads to a decrease in hazardous drinking following the visit. Reinforcement and enhancement of this motivation is one of the likely reasons why structured brief interventions are often more successful than simple advice or feedback. At-risk, rather than dependent, drinkers are also more likely to benefit from brief intervention (Academic ED SBIRT Research Collaborative, 2007; Bazargan-Hejazi et al., 2005; Mello et al., 2005), likely because much of their motivation to change comes from the new realization that their drinking behavior has potential for negative consequences. In several studies, a follow up interaction has been shown to enhance the effectiveness of the brief intervention (Baird et al., 2007; Longabaugh et al., 2001; Mello et al., 2005). This may be due to the fact that the initial intervention leads to further introspection and enhanced motivation to change, which can be reinforced and supported at this follow-up session.

As much potential as screening and ED-based brief intervention seems to offer to change drinking behavior, there are barriers to the success of such programs. Compliance by ED staff with screening and intervention can be poor, most often due to time constraints and demands on staffing (Dent et al., 2008; Graham et al., 2000). To minimize the additional time demands, several studies have investigated the potential for computerized screening and intervention. Computer-based screening has been found to be feasible and effective (Blow et al., 2006; Neumann et al., 2006; Rhodes et al., 2001), thus allowing staff to target their time for brief intervention to screen positive patients. This method also has the benefit of using classically unproductive "waiting time" for active patient assessment. However, the intervention itself appears to require the one-on-one interaction to be effective. Although some benefit has been seen in the adult population, computer-based interventions have not had the success in younger patients that has been seen with personalized brief intervention (Maio et al., 2005).

CONCLUSION

Although primary prevention of alcohol-related injury or illness is still the ideal, adolescents and college-aged students who present to the emergency department are potentially in an enhanced state of receptiveness to intervention and readiness to change their drinking behavior. Through screening and brief intervention which occurs right at the time of their presentation, there is a potential to significantly reduce future hazardous drinking and its associated risk of injury, illness, and even death. Despite the challenges inherent in implementing such programs, the success that has been seen at follow up has been very encouraging and offers a second chance for many youth who are currently at high risk.

Chapter 11

BEHAVIORAL CONTROL MODELS AND ALCOHOL: WHY LOSS OF CONTROL OVER DRINKING MIGHT STEM FROM NEUROCOGNITIVE CHANGES WHEN DRINKING

Lack of self-control and binge drinking appear to be inextricably linked. When an individual engages in binge drinking, any evidence of control over thought and action seems to be lost as the individual continues to drink far beyond the threshold of what is safe to the individual and to others around him or her. Furthermore, the same lack of self-control when binge drinking is evidenced by the increased likelihood that the individual will engage in significantly risky behavior - behavior that, if the person had been sober, would not have been something the individual would consider. Deciding to drive home, despite impairment by alcohol and the risk of being charged by police with impaired driving is one good example. Getting into fights with a stranger at a bar over some minor altercation that would have been ignored if sober is another. Making rude comments to strangers, smoking cigarettes, engaging in risky sex or trying other illegal drugs are also behaviors that might be engaged in when binge drinking, but not engaged in under normal circumstances.

SELF-CONTROL

What is self-control and why is control hijacked when drinking? A challenge for scientists has been to understand how alcohol reduces impulse control in individuals. In general, acute alcohol intoxication is associated with a myriad of negative outcomes, including violent crimes, injuries, and automobile accidents (Pernanen, 1976). Impaired self-control has long been associated with acute alcohol intoxication, and is often a significant predictor of such negative consequences.

In recent years, much has been learned about the acute effects of alcohol on the specific neurocognitive mechanisms that regulate behavioral control. Some of these developments have come from studies of social drinkers in the laboratory (for a review, see Fillmore, 2003). Such research is based on theories that two distinct processes in the brain govern behavioral control (see Figure 11-1).

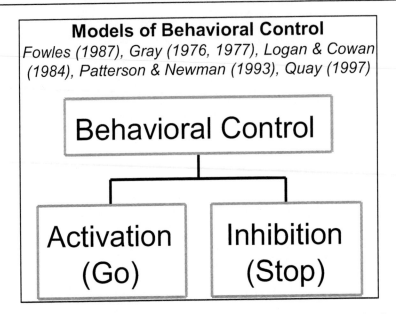

Figure 11.1. The two-process model of behavioral control whereby behavioral control reflects a balance between an activational process and an inhibitory process. If activation dominates, a response will be executed. If inhibition dominates, a response will be suppressed.

One brain process is thought to activate behavior (also sometimes called a 'go process') and the other process is thought to inhibit behavior (also sometimes called a 'stop process') (Fowles, 1987; Gray, 1976, 1977; Logan and Cowan, 1984; Patterson and Newman, 1993; Quay, 1977). In this model, these two processes (activation and inhibition) act in opposition to one another. Thus, the relative strength of each process is assumed to determine behavioral control or self-control. This model assumes that when you observe a person acting impulsively, the process that is thought to inhibit behavior is deficient (Logan, Cowan and Davis, 1984). Deficient behavioral inhibition is considered to be a primary mechanism by which alcohol and other drugs of abuse impair self-control (Fillmore, 2003).

In recent years, model-based assessments of behavioral inhibition mechanisms have been used to study the acute effects of alcohol on the ability to inhibit inappropriate behavioral responses (Fillmore, 2003). Inhibitory processes normally serve to regulate behavior by suppressing or terminating prepotent (i.e., environmentally-triggered) responses (Jentsch and Taylor, 1999). Thus, failures of inhibition result in impulsive and inappropriate actions (Fillmore, 2003). For example, an individual sitting in a bar may want another drink. The sight of alcohol is an environmental stimulus that elicits the wish to drink in an individual. However, if the individual needs to get home since he has to work the next morning, the inhibitory process must suppress the prepotent (i.e., environmentally-triggered) response of taking another drink. If inhibition dominates behavioral control, the response of taking another drink is successfully suppressed and the individual will go home. However, if the individual takes another drink, inhibition has failed, resulting in this impulsive and inappropriate action.

Several studies have examined the effects of alcohol on inhibitory processes using computerized cognitive tasks in the laboratory. By bringing a subject into the laboratory, the researcher can control all extraneous variables that normally impact behavior in the real world. In the above example, the environment is complex. The individual may have friends telling the drinker to stay and have another drink. The bar may be fun and rowdy so that the person wants to stay. A spouse may call the drinker on the phone to tell him to get home now. The amount of alcohol the drinker has already consumed will also matter a great deal as well. However, in the laboratory, these variables can be controlled. A participant in a study can be given a specific dose of alcohol and then tested on a computer task that measures behavioral control. Common cognitive tasks used to assess inhibitory processing include the stop-signal and cued go/no-go tasks. These tasks are laboratory assessments of behavioral control, measuring the ability to quickly activate and suddenly inhibit prepotent responses (Abroms et al., 2003; de Wit et al., 2000, Fillmore and Vogel-Sprott, 1999; Marczinski and Fillmore, 2003a, 2003b, 2005a, 2005b; Mulvihill et al., 1997). Thus, the measurement of response activation involves the measurement of how quickly an individual can make a response, and the measurement of response inhibition involves the measurement of how successfully an individual can withhold making any action.

LABORATORY ASSESSMENT OF BEHAVIORAL CONTROL

If behavior is assumed to be governed by two distinct systems in the brain, where one activates behavior and the other inhibits behavior, then extreme or disinhibited behavior resulting from alcohol consumption could possibly result from one of two scenarios. The first scenario is that an acute dose of alcohol weakens the inhibitory system, causing the disinhibited behavior. The second scenario is that the alcohol heightens the activity of the activation system, also causing the disinhibited behavior. Cognitive neuroscience research has focused much effort on understanding this distinction, and it seems that it is impairment in the inhibitory system which underlies many of the deficits of self-control that have been observed. Such an outcome may not be surprising considering that alcohol is a depressant drug. Thus, it is more likely that an acute dose of alcohol should decrease the effectiveness of a process in the brain (decreased inhibition) instead of increasing the effectiveness of a process in the brain (increased activation).

Initially, the research into understanding how self-control occurs relied heavily on the use of the "stop-signal" model of behavioral control (Logan and Cowan, 1984). As control is determined by competitive activating and inhibiting processes, the researcher could design a computer task to elicit each process by presenting an environmental signal to the participant that requires a response. Thus, 'go' and 'stop' signals, in the form of visual and auditory stimuli such as letters on a computer screen or auditory beeps coming from a computer speaker, are used to elicit these activating and inhibitory processes. For example, a participant may be instructed to press a key on the keyboard every time he sees a letter X or O show up on the computer screen. However, if the participant hears a beep, the participant is not to respond. Thus, the X and O stimuli function as 'go' signals and the beep functions as a 'stop' signal. While the task initially appears quite simple, the participant finds it challenging on some trials. During each trial, the letters are presented. However, on about a ¼ of the trials, the beep is also presented shortly after the presentation of the letter (approximately 50 milliseconds to 300 milliseconds after the onset of the letter). Therefore, on most trials, the participant responds to the letter but on a ¼ of trials, the participant has to withhold his or her response. The reliable presentation of the go stimulus (the letter on the computer screen) on every trial is set up to produce a prepotent instigation to respond. The difficulty in this task is that subjects must overcome the prepotent tendency to respond in order to suppress responses when the stop stimulus (the beep) is presented.

Several aspects of performance can be recorded from results of this stop-signal task. The researcher can measure the participants' mean reaction time to respond to the letters. This reaction time measure is said to be indicative of the state of response activation. If a participant has faster reaction times, greater activation is said to be present. Similarly, the researcher can look at the proportion of trials that the participant successfully withholds a response when the beep is presented. This accuracy measure is said to be indicative of the state of response inhibition. If the participant successfully withholds many responses on those trials when a beep is presented, greater inhibition is said to be present. Other aspects of performance can also be measured using this task, including an estimation of how long the stop signal (beep) can be delayed after the go signal (letter) before the subject can no longer inhibit a response (Logan, 1994).

The important advance of the stop-signal model and its use in understanding self-control is that inhibiting a response is an active cognitive process. It takes a fair amount of cognitive resources to not make a response, particularly when another signal in the environment is cuing one to act. Thus, doing nothing sometimes requires a great deal of cognitive effort. This model has been very influential in our understanding of failures of self-control. For example, attention-deficit hyperactivity disorder (ADHD) is a self-control disorder whereby children have trouble not acting out and sitting still. When children diagnosed with ADHD are compared with control children on the stop-signal task, children with ADHD exhibit impaired inhibitory control (i.e., they have great difficulty in withholding responses to the beeps) even though their activation process appears similar to control children (i.e., they can respond quickly to the letters) (Oosterlaan and Sergeant, 1996; Schachar et al., 1995).

Since impaired self-control is a hallmark feature of alcohol intoxication, several researchers have tested healthy adult participants in the laboratory on the stop-signal task following the administration of a moderate dose of alcohol (de Wit et al., 2000; Easdon and Vogel-Sprott, 2000; Fillmore and Blackburn, 2002; Fillmore and Vogel-Sprott, 1999, 2000; Mulvihill et al., 1997). The results of these studies demonstrated that a moderate dose of alcohol selectively reduced the participants' ability to inhibit their behavior while leaving their ability to activate behavior intact. Thus, a participant who is intoxicated would still be able to quickly and accurately respond to the letters ('go' signals) presented on the computer screen by pressing the appropriate key on the keyboard. However, the participant would not be able to successfully withhold the action when the beep ('stop' signal) was presented. The doses of alcohol typically used to demonstrate these effects were not excessive. For example, Fillmore and Vogel-Sprott (1999) demonstrated this finding in participants who had an average peak blood alcohol concentration of .07g%, which is below the legally sanctioned limit of .08g% used to prosecute drunk drivers throughout North America. Thus, inhibitory mechanisms of behavioral control appear to be particularly sensitive to the impairing effects of alcohol, even at blood alcohol concentrations below the level that society typically considers being intoxicated.

ENVIRONMENTAL CUES AND BEHAVIORAL CONTROL

While the above-described stop-signal task demonstrates that alcohol can impair inhibitory mechanisms of behavioral control, the model does not confer any information about the role that environmental conditions might play in exacerbating or ameliorating the effects of alcohol on behavioral control. Environmental context can be an important determinant in the overall drug effect observed (Falk and Feingold, 1987). For example, aggressive behavior while drinking may be common in a rowdy bar but unlikely at a quiet dinner party, even though the amount of alcohol consumed might be similar. It is also likely that the environment exerts some stimulus control over both the inhibitory and activational processes of behavioral control. For instance, stimulus cues that precede the actual signals to respond or inhibit a response can provide preliminary information about the response that might be required. Using driving as an example, a warning sign of 'new stop sign ahead' can be helpful in warning a driver of a new environmental stimulus requiring a certain response (taking your foot off the gas and pressing it on the brake pedal when the actual stop sign becomes visible). Cognitive psychologists have demonstrated that such predictive cues can significantly facilitate the activation or inhibition of behavior because individuals typically initiate preparatory processes required for the activation or inhibition of an action (Duncan, 1981; Posner, 1980; Posner et al., 1980).

The effects of preliminary cues on the activation and inhibition of behavior have been studied using cued go no-go tasks (Miller et al., 1991). Similar to the above-described stop-signal task, the cued go no-go task measures behavioral control by examining the ability to quickly execute responses (a measure of the activation process) and suddenly suppress responses (a measure of the inhibition process). The difference between the stop-signal task and the cued go no-go task is the inclusion of a manipulation of response prepotency in the cued go no-go task. In other words, in the cued go no-go task, a preliminary 'go' or 'no-go' cue is presented before the actual go or no-go target is displayed. The cue provides some information about the likelihood that a go or no-go target will be presented. The cue-target relationship is manipulated so that cues have a high probability (usually 80%) of correctly signaling a go or no-go target. There is a low probability (usually 20%) that a cue will incorrectly signal a target. By virtue of this cue-target relationship, correct (valid) cues facilitate response activation and response inhibition.

In a typical cued go no-go task, the participant watches a computer screen and stimuli are presented to the participant. The cue is a rectangle that is visually presented in either a horizontal or vertical orientation. The orientation of the cue (horizontal or vertical) signals the probability that a go or no-go target would be displayed. The go and no-go targets are the colors green and blue, respectively. The colors are presented as solid hues that fill the interior of the rectangle cue. The participant is asked to press a key on the keyboard every time a green rectangle appears. The participant is told to not make a response every time a blue rectangle appears. The participant is not given any information about the empty rectangles except that they are not to respond to them and they help you get ready for a color that may appear. Figure 11-2 illustrates a typical trial sequence in a cued go no-go task. A participant would usually perform 250 trials of which half are go trials (a green rectangle is presented) and the other half are no-go trials (a blue rectangle is presented). Figure 11-3 illustrates that proportion manipulation is used to induce response prepotency.

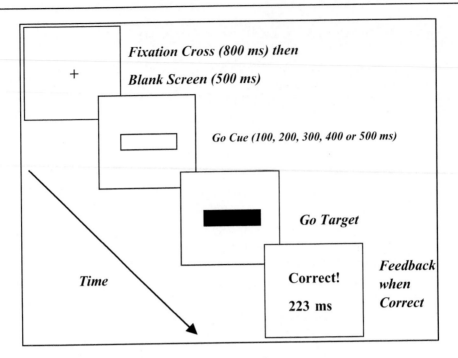

Figure 11.2. A typical trial sequence in a cued go no-go task.

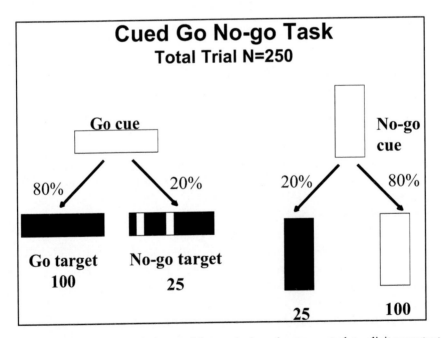

Figure 11.3. The proportion manipulation used in a typical cued go no-go task to elicit prepotent responding.

Cues are valid on 80% of trials and invalid on 20% of trials. Therefore, the subject relies on the cues to predict what target is likely to appear (green or blue). This task takes about 15 minutes to complete (Marczinski et al., 2007; Marczinski and Fillmore, 2003a, 2003b, 2005a, 2005b).

Participants will quickly learn the relationship between the cues (horizontal or vertical empty rectangle) and targets (green or blue rectangle). Therefore, performance on this task demonstrates that presentation of a valid go cue before the actual presentation of a go target typically results in faster reaction times to the go target. This speeding effect due to the valid cue is attributed to covert response preparation that occurs before the actual go target is presented (Posner, 1980). Similarly, accurate suppression of responses is more likely when a valid no-go cue is presented before the actual presentation of a no-go target. This further illustrates the fact that inhibition is an active cognitive process. Preparation in response to not act can be as important as preparation to make an action. However, the speeding of responses to go-targets and increased accuracy in suppressing responses to no-go targets owing to the reliance on preresponse cues to guide behavior comes at a cost on the minority of trials when cues are inaccurate in its prediction (i.e., when a go cue is presented before a no-go target or a no-go cue is presented before a go target). Typically, responses to go targets are significantly slower following no-go cues and failures to suppress responses to no-go targets are more frequent following go cues (Miller et al., 1991).

The acute effects of alcohol on activational and inhibitory processes have been examined using this cued no-go task (Marczinski and Fillmore, 2003a, 2003b, 2005). The effects of alcohol on the inhibition and activation of responses were dependent on the cues that signaled the likelihood that a response should be executed or suppressed (see Figures 11-4 and 11-5).

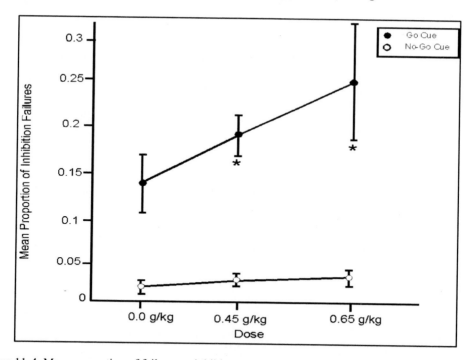

Figure 11-4. Mean proportion of failures to inhibit responses to no-go targets following invalid go and valid no-go cues under three alcohol doses: 0.0 g/kg (placebo), 0.45 g/kg and 0.65 g/kg (see Marczinski and Fillmore, 2003b). Capped vertical lines show standard errors of the mean.

Alcohol produced a dose-dependent increase in inhibitory failures following invalid go cues but had no effect on inhibitory failures following the valid no-go cues. The same cue-dependent pattern was also observed for response execution. Alcohol slowed mean reaction times in a dose-dependent manner following the invalid go cues, but it had no effect on reaction time following the valid go cues.

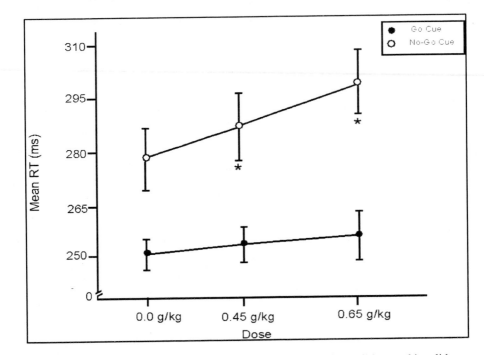

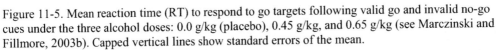

Figure 11-5. Mean reaction time (RT) to respond to go targets following valid go and invalid no-go cues under the three alcohol doses: 0.0 g/kg (placebo), 0.45 g/kg, and 0.65 g/kg (see Marczinski and Fillmore, 2003b). Capped vertical lines show standard errors of the mean.

Thus, under alcohol, participants rely more heavily on environmental cues to guide behavior. This helps maintain some behavioral control if the cues are valid. However, behavioral control is quite impaired if the environmental cues are invalid. For example, think back to the above example of the individual sitting at the bar pondering whether to have another drink. If the environmental cues favor inhibition (e.g., a friend says that we had better leave because you have to work tomorrow), then behavioral control over drinking is maintained. However, if the environmental cues favor activation (e.g., a friend says that we should have another drink and stay), then behavioral control over drinking may be lost. Even in a sober state, there will always be internal sources and external sources that contribute to behavioral control. However, when drinking, the source of control over behavior may transition from greater internal sources to greater external sources.

Studies that have used the cued go no-go task have found that it only requires moderate doses of alcohol to significantly impair the ability to inhibit a prepotent (i.e., instigated) response. The impairing effects of alcohol were evident in spite of the relatively simple nature of the inhibitory response tested in these models. All subjects are required to do is push a key on a keyboard when they see a green rectangle and do nothing when they see a blue rectangle. Similarly, the impairing effects of alcohol were evident even though the doses of alcohol that were administered were comparatively mild. The doses of alcohol used in these studies resulted in blood alcohol concentrations in the range of .05 to .08 g% (i.e., below or at the legal limit for driving).

ALCOHOL IMPAIRMENT OF INHIBITORY CONTROL AND ACTUAL DRINKING

Since moderate doses of alcohol impair the ability to inhibit a response, this increase in impulsivity may contribute to binge drinking and alcohol abuse. The alcohol-induced impairment of inhibitory control could compromise the drinker's ability to stop the self-administration of alcohol, thus increasing the amount of alcohol consumed within one episode. Weafer and Fillmore (2008) recently examined the relationship between acute alcohol impairment of inhibitory control and alcohol consumption during a single drinking episode in the laboratory. Participants were recruited for a three session study. The authors first brought college students into the laboratory to measure the degree to which their inhibitory control was impaired by a moderate dose of alcohol (0.65 g/kg) versus placebo. For the first two sessions, each participant was given a drink (either alcohol or placebo) and then administered the cued go/no-go task to measure inhibitory control. Degree of impairment of inhibitory control under alcohol compared to placebo was calculated for each participant.

On the third session, participants' alcohol consumption was assessed in the laboratory. To do this, participants were told that they would complete a taste-rating task. Participants sampled four different beers and rated them on various qualities, ostensibly to provide information on beer preferences in order to aid in future research. The tasting portion lasted 90 minutes and participants were told that they may drink as much or as little of each beer as they liked, but to be sure to sample enough of each beer to give an accurate rating. Following the taste rating, participants' blood alcohol concentrations were measured and the amount of alcohol consumed by each participant was recorded. Fillmore and Weafer (2008) found that the magnitude of impairment of inhibitory control from alcohol (measured on sessions one or two) was associated with increased consumption of alcohol (on session three). Thus, acute impairment of inhibitory control might be an important cognitive effect that contributes to binge drinking, in addition to the positive rewarding effects of alcohol.

BINGE DRINKERS AND BEHAVIORAL CONTROL

Binge drinkers appear to significantly lose behavioral control when drinking. Binge drinkers are 14 times more likely to drive while impaired by alcohol than are non-binge drinkers (Naimi et al., 2003). Binge drinking is also associated with numerous other behaviors that demonstrate loss of control, including unplanned and unsafe sexual activity, assaults, falls, injuries, criminal violations, and automobile crashes (Wechsler et al., 1994, 1995, 1998, 2000). One potential explanation for the correlation between binge drinking and elevated alcohol-related behaviors that demonstrate loss of control is that binge drinkers are more impulsive in general and are more disinhibited by alcohol compared to more moderate drinkers.

The hypothesis has been tested that binge drinkers would suffer greater impairments in behavioral control in response to a moderate dose of alcohol compared to their non-binge drinking peers. Thirty-two young college student social drinkers of equal gender (16 males and 16 females) were selected to participate in this study and were classified as binge and non-binge drinkers using the well-established Wechsler definition of binge drinking. With this definition, binge drinking was defined as the consumption of five or more drinks per episode for men and four or more drinks per episode for women (Wechsler et al., 1994, 1998, 2000). This measure assumes that drinking is to intoxication if the individual meets or exceeds the five/four drinks criterion. Eighteen participants were considered to be binge drinkers and 14 were considered to be non-binge drinkers. Participants attended a session during which they received a moderate dose of alcohol (0.65 g/kg) and a session during which they received a placebo. In the alcohol session, the mean peak blood alcohol concentration was .085g%, which is just above the legal limit for driving. Participants' inhibitory control in response to each dose using the cued go/no-go task was measured.

Figure 11-6 illustrates the failures to inhibit responses to the no-go targets following the invalid go cue (the measure known to be sensitive to the impairing effects of alcohol) under alcohol and placebo for both the binge and non-binge drinkers. Compared with placebo performance, alcohol increased inhibition failures in the binge drinkers. By contrast, inhibition failures in the non-binge drinkers did not differ between the alcohol and placebo conditions. Interestingly, binge and non-binge drinkers did not differ in their level of inhibition failures following placebo administration. Thus, group differences in disinhibition were only observed when a moderate dose of alcohol was administered. There were no differences between the groups on the reaction time measure of activation (see Marczinski et al., 2007).

Participants were also asked how they felt after receiving the alcohol or placebo dose by having the participants complete a questionnaire (the Biphasic Alcohol Effects Scale) that asked them to rate how stimulated and sedated they perceived themselves to be. This questionnaire lists 14 adjectives and asks the participant to rate the degree to which drinking produced each feeling on a 11-point Likert-type scale, ranging from 0 (not at all) to 10 (extremely). There are seven adjectives that describe stimulation effects (e.g., stimulated, elated) and seven adjectives that describe sedation effects (e.g., sedated, sluggish). The stimulation and sedation scores were summed separately to provide a total subscale score for stimulation and sedation (score range 0-70).

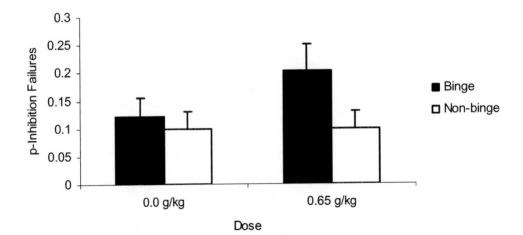

Figure 11-6. Mean proportion of failures to inhibit responses to no-go targets following invalid cues under the 0.0 g/kg (placebo) and 0.65 g/kg alcohol dose conditions for binge and non-binge drinkers. Vertical bars show standard errors of the mean. See Marczinski et al. (2007) for more details.

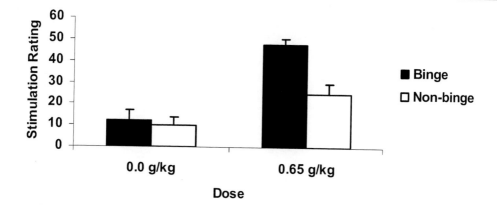

Figure 11-7. Mean self-reported subjective responses of stimulation from the Biphasic Alcohol Effects Scale under the 0.0 g/kg (placebo) and 0.65 g/kg alcohol dose conditions for the binge and non-binge drinkers. Vertical bars show standard errors of the mean. See Marczinski et al. (2007) for more details.

Figure 11-7 illustrates how the ratings of stimulation did not differ between the binge and non-binge drinkers following administration of the placebo. By contrast, the binge drinkers reported feeling much more stimulated by the active dose of alcohol compared to the non-binge drinkers. Thus, the binge drinkers were more disinhibited (as measured by the cued go no-go task) and reported feeling more stimulated (as measured by the Biphasic Alcohol Effects scale) under alcohol compared to their moderate drinking peers.

The finding of a heightened disinhibitory reaction to alcohol in binge drinkers may help to explain the link between impulsive behaviors and problem drinking at a more fundamental level of behavioral control. The heightened disinhibitory response to alcohol might reflect greater vulnerability of inhibitory control mechanisms to alcohol. Such vulnerability in inhibitory functioning could compromise the ability to stop additional alcohol consumption in a drinking situation, leading to binge drinking. Greater disinhibition under alcohol might also explain the increased incidences of other disinhibited behaviors (e.g., unsafe sexual activities, assaults, etc.) in binge drinkers.

An intriguing aspect of this finding of a heightened disinhibitory reaction to alcohol in binge drinkers is that that the finding runs counter to what would be predicted from basic principles of pharmacological tolerance. While tolerance will be discussed in greater detail in chapter 14, it is worth mentioning briefly here. The binge drinkers in the Marczinski et al. (2007) study reported higher levels of alcohol consumption per episode and drank more frequently. From a pharmacological perspective, these individuals would be expected to be more tolerant to the acute effects of the alcohol dose since their body should be more familiar with that amount of alcohol. However, the binge drinkers actually displayed more impairment in inhibitory control and greater increases in stimulation following alcohol consumption. Little is known about the actual behavioral tolerance of binge drinkers and whether it is uniform across various domains of behavioral functioning. Perhaps binge drinking does not afford any appreciable tolerance to inhibitory mechanisms in the brain.

What is still unknown is why binge drinkers have this heightened disinhibitory reaction to alcohol and more moderate drinkers do not? There is currently no hypothesis to explain whether the heightened reaction stems from the increased alcohol consumption associated with binge alcohol use causes inhibitory mechanisms in the brain to decrease in effectiveness over time. Alternatively, binge drinkers may be predisposed to disinhibition before they ever consume their first drink of alcohol. If this is the case, drinkers' disinhibitory response to alcohol may prove to be an important indicator of current or future alcohol-related problems. Until studies are completed that further investigate the heightened disinhibitory reaction with individuals who have differing histories of binge alcohol use, it will remain a challenge to disentangle what causes this problem and what it means for future problems with alcohol.

CONCLUSION

Stop-signal and cued go/no-go tasks have provided important insights into the specific brain mechanisms by which alcohol appears to impair self-control. Various research studies of alcohol effects using these tasks have provided compelling evidence that alcohol reduces self-control by impairing the response inhibition process in the brain. When response inhibition is impaired, activation is left to dominate behavior. Overactive and impulsive behavior is the result. This problem of alcohol-induced impairment of inhibition appears to be significantly worse for binge drinkers. Compared to their more moderate drinking peers, binge drinkers are more disinhibited by an acute dose of alcohol on the cued go/no-go task. These group differences in disinhibition were only observed when a moderate dose of alcohol was administered. Thus, the heightened disinhibitory reaction to alcohol in binge drinkers may help explain the link between impulsivity and problem drinking at the fundamental level of behavioral control. This vulnerability in inhibitory control mechanisms could compromise the ability to stop additional alcohol consumption in a drinking context, leading to binge drinking. Furthermore, the vulnerability of inhibitory control mechanisms to alcohol may also explain the relationship between binge drinking and other self-control failures such as unplanned and unsafe sexual activity, assaults, fall, injuries, criminal violations and impaired driving.

Chapter 12

BINGE DRINKERS AND THEIR PARTICULAR VULNERABILITY TO THE EFFECTS OF ALCOHOL

Binge drinkers appear to have a particular vulnerability to the disinhibiting effects of alcohol. Many behavioral problems associated with binge drinking, such as aggressive behavior and poor decision making, may stem from the basic problem of vulnerability to the disinhibiting effects of alcohol. In the previous chapter, the laboratory evidence supporting the model that alcohol impairs self-control by causing neurocognitive changes was introduced. Once alcohol is consumed, the inhibitory processes in the brain that normally maintain self-control become less effective. For example, an individual may have excellent self-control when sober. However, after a few drinks, impulse control is compromised and the individual may continue to drink alcohol, make bad decisions or act aggressively. It is the decreased impulse control that may lead to a binge episode as the initial consumption of some alcohol decreases control over consumption and the individual continues to drink far beyond what may have been the original intent. Alcohol-induced disinhibition also impacts other behaviors and can cause widespread problems (see Figure 12-1).

A good real-life example of the behavioral effects of alcohol-induced disinhibition is reflected in the details of the arrest of actor-director Mel Gibson in 2006. Mel Gibson was arrested for drunk driving with a blood alcohol concentration of approximately .12 g%, an amount that is well above the legal limit of .08 g% in California. While many Hollywood actors have been arrested for impaired driving, the Gibson arrest made international news because he purportedly made several anti-Semitic statements to police during the arrest (Maugh, 2006). Following the logic of the model that alcohol impairs inhibitory control, it may be possible that Mel Gibson's anti-Semitic beliefs are normally held in check when he is sober. However, his anger at being arrested and his level of intoxication may have contributed to the outburst, possibly revealing impaired inhibitory control and resulting in statements that normally would be held in check. Similarly, the disinhibitory effects of alcohol may also have led to Gibson's initial decision to drive despite knowledge that this behavior is not only dangerous but illegal and might have negative effects for him professionally and personally.

Binge drinkers have more impairment of inhibitory control under the influence of alcohol compared to their moderate drinking peers (Marczinski et al., 2007). Given the same dose of alcohol, young college students who binge drink exhibit greater loss of impulse control compared to their peers who don't consume alcohol in this manner.

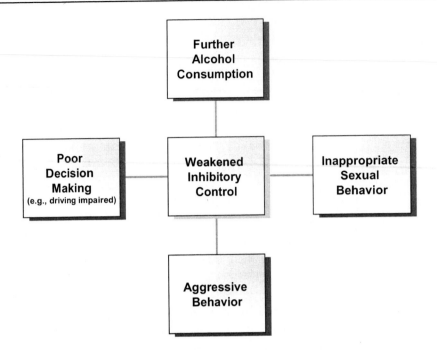

Figure 12.1. Weakened inhibitory control may have widespread effects on other behaviors including heightened aggressive behavior, inappropriate sexual behavior and a further escalation of drinking.

While it is unknown if alcohol-related impairment of inhibitory control is a cause or consequence of binge drinking (as there are no published studies that have compared inhibitory control in binge and non-binge drinkers before initiation of drinking), this altered neurocognitive functioning in binge drinkers will fundamentally alter other important behaviors. In this chapter, the effects of binge drinking on alcohol-induced impairment of other behaviors are examined, including behaviors such as driving, aggression, and the further consumption of alcohol.

BINGE DRINKERS, ALCOHOL, AND DRIVING

Alcohol remains one the principal risk factors for motor vehicle crashes (NIDA, 1985). Binge drinking in particular seems to be associated with impaired driving, with binge drinkers being 14 times more likely to drive while impaired by alcohol compared with non-binge drinkers (Naimi et al., 2003). Driving while intoxicated is more significantly associated with binge drinking than with other high risk drinking behaviors such as chronic heavy drinking (Borges and Hansen, 1993; Duncan, 1997). It has been estimated that in over 80% of episodes of alcohol-impaired driving, the individuals were engaging in binge drinking (Quinlan et al., 2005). In 2002, there were over 17,000 motor vehicle fatalities in the United States involving alcohol, representing an average of one alcohol-related fatality every 30 minutes. The highest intoxication rates of fatal crashes were recorded for drivers between the ages of 21 and 24 (NHTSA, 2002) which also is the age group who is most likely to binge drink (Wechsler et al., 1994, 1995, 1998, 2000). Thus, college students, the demographic group with the highest rates of binge drinking, are also particularly prone to alcohol-impaired driving. A national survey of self-reported behavior in the past 30 days reported that 3 in 10 college students reported driving after drinking any amount of alcohol and 1 in 10 students reported driving after consuming five or more drinks (Wechsler et al., 2003). Other surveys have estimated that in any given year, over 2 million of the 8 million college students in the U.S. drove under the influence of alcohol at least once (Hingson et al., 2002). Despite extensive alcohol education programs on college campuses and in the general public and increasingly strict legal ramifications of this behavior, the underlying reasons for the high rates of impaired driving in college age binge drinkers are not well understood.

A recent study may provide some indications as to why impaired driving rates are so high in college age binge drinkers. Marczinski et al. (2008) studied the acute effects of alcohol on driving performance (using a driving simulator in the laboratory) and subjective ratings of intoxication and driving ability in binge and non-binge drinkers (Marczinski et al., 2008). Young social drinking college students (24 binge drinkers and 16 non-binge drinkers) were brought in to the laboratory for two different sessions. In one session the participant received a moderate dose of alcohol (0.65 g/kg) and on another session the participant received a placebo (a drink that looked and smelled like alcohol, but contained no alcohol). In the alcohol session, the mean peak blood alcohol concentration was .09g%, which is just above the legal limit for driving. Binge and non-binge drinkers were classified using the well-established Wechsler definition of binge drinking, with binge drinking defined as the consumption of five or more drinks per episode for men and four or more drinks per episode for women (Wechsler et al., 1994, 1998, 2000). This measure assumes that drinking is to intoxication if the individual meets or exceeds the five/four drinks criterion. All participants were tested on the driving simulator in response to each dose and were also asked to give subjective ratings of their reaction to the dose, including if they felt able to drive.

The driving simulator provided a realistic experience of driving for the participants. Each individual was seated in front of a computer display that provided a full view of the roadway and vehicle instrument panel (e.g., speedometer). The participant was instructed to control the vehicle down the road by moving the steering wheel and manipulating the accelerator and brake pedals. The participant was asked to accelerate and maintain a constant speed of 55 mph while remaining in the center of the right lane of the roadway (lane width of 12 feet). The 20-minute driving course covered approximately 19 miles and included a winding road, hills and building by the side of the road. During the drive, other passing vehicles were present but they did not require the participant to pass them or brake for them. The simulation presented clear weather daytime lighting conditions. While this simulation involved a simple drive, previous studies have shown this drive to be sensitive to the impairing effects of moderate doses of alcohol (Harrison and Fillmore, 2005; Harrison et al., 2007).

Figure 12-2 illustrates the results for the within lane deviation measure (a measure of swerving) from the 20 minute driving simulation under alcohol and placebo for both the binge and non-binge drinkers. Compared with placebo performance, alcohol increased swerving in both groups. Both binge and non-binge drinkers appeared to be similar in their performance on the within-lane deviation measure under both dose conditions. Binge and non-binge drinkers were also impaired by alcohol on other aspects of the simulated driving task. Under alcohol, both groups also had more difficulty maintaining the required 55 mph speed as speed deviation increased under the alcohol dose. Both groups also exhibited increased driving errors under alcohol including crossing over the center line (into oncoming traffic), driving over the shoulder and having more accidents. Thus, both binge and non-binge drinkers were similarly impaired by alcohol on multiple aspects of the simulated driving task compared to placebo performance.

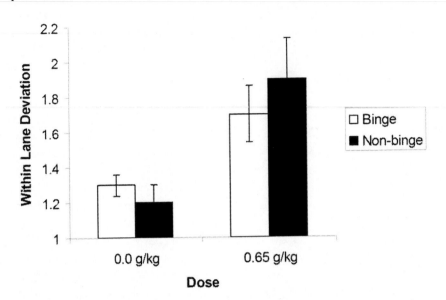

Figure 12.2. Within-lane deviation (in feet) under the 0.0 g/kg (placebo) and 0.65 g/kg alcohol dose conditions for binge and non-binge drinkers. Vertical bars show standard errors of the mean. See Marczinski et al. (2008) for more details.

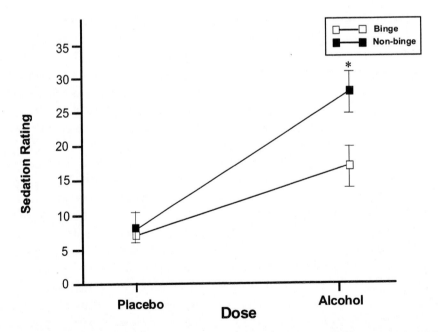

Figure 12.3. Mean self-reported subjective responses of sedation from the Biphasic Alcohol Effects Scale under the 0.0 g/kg (placebo) and 0.65 g/kg alcohol dose conditions for the binge and non-binge drinkers. Vertical bars show standard errors of the mean. *Significant difference between binge and non-binge drinkers ($p < .05$).

Contrasting with the simulated driving performance data, binge and non-binge drinkers had very different subjective reactions to the alcohol. First, participants were asked the rate how they felt after being given the alcohol or placebo dose by filling out a questionnaire (the Biphasic Alcohol Effects Scale) that asked them to rate how stimulated and sedated they perceived themselves to be (Martin et al., 1993). This questionnaire listed 14-adjectives and asked the participants to rate the degree to which drinking produced each feeling on a 11-point Likert type scale, ranging from 0 (not at all) to 10 (extremely). There were seven adjectives that described stimulation effects (e.g., stimulated, elated) and seven adjectives that described sedation effects (e.g., sedated, sluggish). The stimulation and sedation scores were summed separately to provide a total subscale score for stimulation and sedation (score range 0-70). Figure 12-3 illustrates how the ratings of sedation did not differ between the binge and non-binge drinkers following administration of the placebo. By contrast, the binge drinkers reported feeling much less sedated by the active dose of alcohol compared to the non-binge drinkers.

Participants were also asked to complete an able-to-drive rating (Beirness, 1987). The able-to-drive rating is a visual analogue scale that asks participants to rate how able they think they are to drive an actual vehicle at the time they are completing the questionnaire. Participants indicate their response ranging from 'not at all' to 'very much' by placing a vertical mark through a 100-mm line. Figure 12-3 illustrates how the ability to drive ratings did not differ between binge and non-binge drinkers following administration of the placebo. By contrast, the binge drinkers reported that they were much more able to drive following the active dose of alcohol compared to the non-binge drinkers.

The pattern of results from the Marczinski et al. (2008) may explain why binge drinkers are at increased risk for engaging in impaired driving.

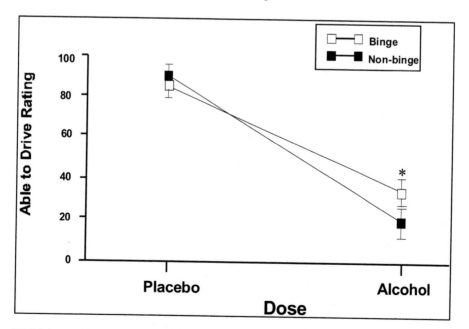

Figure 12-4. Mean self-reported subjective responses of ability to drive under the 0.0 g/kg (placebo) and 0.65 g/kg alcohol dose conditions for the binge and non-binge drinkers. Vertical bars show standard errors of the mean. *Significant difference between binge and non-binge drinkers ($p < .05$).

The binge drinkers reported feeling less sedated by the 0.65 g/kg dose of alcohol compared to the non-binge drinkers. Furthermore, binge drinkers reported greater driving ability under alcohol compared to the non-binge drinkers. Alcohol impaired multiple aspects of simulated driving performance in all participants and binge and non-binge drinkers did not differ on any aspect of driving performance following alcohol or placebo administration. Thus, these results indicate that the epidemiological reports of high rates of impaired driving in binge drinkers may be due to a dissociation between behavioral impairment and subjective rating of impairment following consumption of alcohol.

The findings of Marczinski et al. (2008) study coincide with the recent reports that binge drinkers report feeling less impaired by alcohol (Brumback et al., 2007; Holdstock et al., 2000; Marczinski et al., 2007; Rose and Grunsell, 2008) and have a heightened disinhibitory reaction to alcohol (Marczinski et al., 2007) compared to more moderate social drinkers. However, the underlying differences in biological and psychological mechanisms between binge and non-binge drinkers are unknown. It is possible that binge and non-binge drinkers differ in the types of effects that they expect from alcohol, and such expectations could influence their willingness to drive under the drug (Vogel-Sprott and Fillmore, 1999). Alternatively, binge drinkers may be more tolerant to the subjective impairing effects of alcohol, compared to non-binge drinkers. In the Marczinski et al. (2008) study, the binge drinkers typically drank twice as often with twice as many drinks as the non-binge drinkers. Tolerance (the reduced response to a drug following repeated administrations) may have manifested itself in the subjective reactions to alcohol in the binge drinkers. Learning is often an important factor in the development of tolerance, as alcohol tolerance is often environmentally dependent, whereby maximum tolerance is observed when the drug is taken in the presence of predrug cues that reliably signal its administration (Siegel, 1989; Vogel-Sprott and Fillmore, 1999). In other words, a drinker will be more tolerant to alcohol when drinking in the same bar with the same friends as usual, in comparison to alcohol consumption that occurs in a new location with strangers. By virtue of their drinking history, the binge drinkers in the Marczinski et al. (2008) study had many more instances to develop tolerance to the subjective effects of alcohol compared with the non-binge drinkers. By contrast, tolerance to the behavioral impairing effects of alcohol on driving is unlikely to have developed in the binge drinkers since there are (hopefully) few occasions where the individual has had the opportunity to practice driving while intoxicated.

Overall, the Marczinski et al. (2008) study illustrates that binge and non-binge drinkers differ in their subjective perceptions of their level of intoxication, with individuals who habitually binge drink not feeling like they are too intoxicated to drive. These subjective feelings are at odds with their actual driving performance under alcohol, which remains poor. The results of this study also add to the growing body of literature that suggests people tend to be poor estimators of intoxication. In particular, individuals often underestimate their blood alcohol concentration (BAC) and the amount of alcohol they consume (Fillmore et al., 2002; Harrison and Fillmore, 2005b; Ogurzsoff and Vogel-Sprott, 1976; Russ et al., 1986). Judgments regarding the ability to drive legally (i.e., at a BAC below the legal limit) become less accurate as BAC increases (Beirness, 1987). Drinkers also underestimate the time required for BAC to reach peak levels after drinking ceases (Portans et al., 1989). Thus, evidence that drinkers underestimate BACs is an important finding because it could provide some insight into for why individuals might choose to drive after drinking. Drinkers could make erroneous assumptions that their BAC is below the legal limit, and thus decide it is safe to drive. The results of Marczinski et al. (2008) suggest that this problem is exacerbated in binge drinkers compared to their more moderate drinking peers.

BINGE DRINKERS, ALCOHOL AND AGGRESSION

Aggression has been defined as any behavior intended to harm a person who would prefer not to receive such treatment (Bushman and Cooper, 1990). Alcohol intoxication has a high correlation with all types of aggressive behavior, such as criminal activity ranging from property offenses to violent crimes. Violent crimes include murder or attempted murder, manslaughter, rape or sexual assault, robbery, assault and other crimes such as kidnapping, hit-and-run driving, and child abuse. When people commit violent crimes, they tend to be under the influence of alcohol (Collins, 1980). The co-occurrence of violent crime and alcohol use is highly prevalent among men between the ages of 18 and 30, as this demographic group has a relatively high rate of both criminal activity and heavy drinking. In addition, alcohol is involved in a high proportion of spousal abuse incidents with estimates that either the offender or victim was using alcohol in 25-50% of cases (Collins, 1980).

Crime is often linked to binge drinking, with young adults being the demographic group most likely to engage in each of these activities. The results of one survey of individuals between the ages of 12 and 30 years old living in England and Wales confirmed that binge drinkers, and in particular male binge drinkers between the ages of 18 and 24 years, were most likely to engage in criminal behavior. In the 12 months prior to the interview, 39% of binge drinkers admitted to committing at least one offense and 60% of binge drinkers admitted to committing at least one criminal and/or disorderly act during or after drinking alcohol. Getting drunk was highly predictive of aggressive behavior. Individuals who reported getting drunk at least once per week were more than five times more likely to be involved in a fight or other violent crime and had a seven times greater probability of breaking or damaging something (Richardson and Budd, 2003). Similar findings also have been reported for adolescents in the United States. Zhang and Johnson (2005) surveyed a representative sample of 9,058 Mississippi high school students. They reported that males and binge drinkers reported an increased likelihood of violence-related behaviors on school property. While male binge drinkers are most likely to engage in violent crimes, female binge drinkers are at an increased risk of being a victim of a crime. Champion et al. (2004) surveyed adolescent females and found that self-reported binge drinking was associated with sexual victimization (i.e., actual or attempted sex against their will).

While violent crimes are the most disconcerting correlate of binge drinking, other aggressive behaviors have also been reported. For example, Bichler and Tibbetts (2003) surveyed 289 college students about their drinking habits and their self-reported academic dishonesty. They reported that binge drinkers were more likely to report cheating behavior and also were more likely to report low self-control on a questionnaire. Thus, the correlation between binge drinking and aggression appears to be widespread, ranging from the more benign to extreme outcomes.

Disinhibition theory holds that alcohol releases behavior normally kept under self control (such as aggressive impulses or sexual impulses). The strong correlation between alcohol and aggression suggests that alcohol-induced failures of self control may be contributing to a variety of aggressive behaviors. If binge drinkers are more disinhibited by alcohol (Marczinski et al., 2007), then it is not surprising that binge drinkers are also more likely to behave aggressively when drinking. However, research has suggested that there are other important factors at play impacting the relationship between alcohol and aggression. Other variables must be important given the distinct gender difference in the relationship between binge drinking and aggressive behavior. While there are both male and female college students who are binge drinkers, it is almost uniquely males who engage in alcohol-induced aggressive behaviors. It has been shown that alcohol combined with situational factors, such as social pressure and threat of retaliation (Adesso, 1985; Graham et al., 1998), as well as personal factors, such as how angry a person characteristically is (Parrot and Giancola, 2004) have been demonstrated to play a significant role in this relationship.

Other research has demonstrated that drinking context (being at a party or at a fraternity/sorority) combined with the number of drinks consumed was positively associated with aggression (Wells et al., 2008). Thus, alcohol does not simply cause aggression via disinhibition in the brain. Aggression is a complex social behavior affected by the characteristics of the aggressor as well as situational factors, only one of which is alcohol consumption. However, disinhibition by alcohol is most certainly part of the equation of the relationship between alcohol and aggression.

BINGE DRINKERS, ALCOHOL AND RISKY SEXUAL BEHAVIOR

Binge drinking is also strongly associated with risky sexual behaviors. In a case-control study of women with pregnancies that resulted in a live birth, Naimi et al. (2003) compared women with unintended pregnancies with those with intended pregnancies. With an impressive sample size of 72,907 respondents, the authors reported that in the entire sample, 14% of the women reported binge drinking and 45% of the binge drinkers' pregnancies were unintended. Binge drinking in the preconception period was strongly associated with unintended pregnancies.

Women who had unintended pregnancies were also more likely to be younger, unmarried, smokers and reported being exposed to violence. Women binge drinkers were also more likely to consume alcohol, binge drink and smoke during the pregnancy, with all of these factors having known teratogenic effects on the developing fetus.

Guo et al. (2002) provided some insight into the developmental relationship between adolescent binge drinking and risky sexual behavior. The authors followed an urban sample of 808 children from Seattle, Washington. The children were surveyed at age 10 and then followed prospectively up until age 21. The results indicated that adolescents who engaged in binge drinking were also more likely to engage in risky sexual behavior, including having an increased number of sexual partners and a decreased likelihood of using a condom consistently compared to their non-binge drinking peers.

BINGE DRINKERS, ALCOHOL AND PRIMING THE MOTIVATION TO DRINK

In this chapter, several examples of how alcohol-induced disinhibition impacts other behaviors, such as aggression and risky sexual behavior, have been described. However, the most important consequence of alcohol-induced disinhibition may be that this decreased inhibitory control leads to the further consumption of alcohol. As such, a vicious cycle develops. A small amount of alcohol (the equivalent of about two drinks) is all that is needed to result in impaired inhibitory control (Marczinski and Fillmore, 2003a) and could be the starting point to precipitate this cycle (see Figure 12-5). It is commonly assumed that binge drinking occurs because the initial drink primes continued alcohol intake in a particular situation (Chutuape et al., 1994; Ludwig et al., 1974; Marlatt and Gordon, 1980). There are obviously marked individual differences in the priming effects of a small amount of alcohol, as some individuals routinely drink a small amount and stop while others exhibit loss of control over their drinking (Fillmore, 2001).

Research on this vicious cycle between alcohol consumption and disinhibition was initially of interest for the treatment of alcoholism. A common observation of substance abuse counselors was that an abstinent alcoholic could be successfully abstinent for long periods of time only to have one drink that began a downward spiral back into full-blown alcohol dependence. Today, successful treatment programs such as Alcoholics Anonymous (AA) incorporate total abstinence from alcohol as part of their approach to successful recovery. In the first step of the twelve-step program, an individual in AA dismisses the notion that alcoholics can control their drinking or can ever reach that position. Recovery only occurs when the alcoholic admits that he or she is powerless over alcohol, that without alcohol a return to health is possible, and that with alcohol the downward spiral to self-destruction continues (Alcoholics Anonymous, 1972, 1983).

Weakened Inhibitory Control   **Alcohol Consumption**

Figure 12.5. The vicious cycle of a small dose of alcohol leading to greater disinhibition which leads to greater consumption of alcohol. This pattern may be exacerbated to a certain degree in binge drinkers and to a large extent in alcoholics.

In other words, the only way to break the vicious cycle shown in Figure 12-5 is to never touch a drop of alcohol again, which is why an individual who successfully completes AA calls him or herself a recovering alcoholic, even if that person has not exhibited symptoms of alcohol dependence for decades.

However, the importance of total abstinence in recovery has recently become more controversial, as others have argued that a return to moderate drinking (not total abstinence) should be the goal of a treatment program, especially since alcohol consumption is so pervasive in our society. Whether moderate drinking or total abstinence should be the treatment goal depends on the client and severity of the alcohol dependence. A moderate drinking goal may be more appropriate if individuals exhibit a less severe alcohol dependence problem, believe that moderate drinking is possible, are of a younger age, are employed, and are currently living with psychological and social stability (Rosenberg, 1993). Similarly, advocates of harm reduction programs argue that the goal of treatment should be to reduce the negative consequences of using alcohol, rather than on reducing the quantity or frequency of its consumption (Single, 1995).

Controversies over the best way to treat alcohol dependence aside, there is good research evidence that some small amount of alcohol consumed often leads to further consumption of alcohol, which is also true for most drugs of abuse. The ability of an initial drink to positively reinforce additional consumption of alcohol was called the *priming effect* (de Wit, 1996). A researcher can measure priming in several ways. One common methodology to measure priming is to use self-reported ratings of motivation to use alcohol following an initial dose (Chutuape et al., 1994; de Wit and Chutuape, 1993; de Wit and Griffiths, 1991). For example, a mild dose of alcohol or placebo is first administered to social drinkers in the laboratory. The subjective effects of alcohol and the participant's motivation to consume alcohol is then measured. Priming effects of the motivation to drink further amounts of alcohol are assessed by questions such as, "How much desire do you have for alcohol right now?" Ratings are usually provided on visual analogue scales, such as line where the participant rates from not at all at one end to very much at the other end of the line. The typical result is that compared with placebo, the administration of a low dose of alcohol produces priming (i.e., increased desire to consume alcohol) as well as increased subjective effects (e.g., increased liking of the alcohol) in social drinkers. The alcohol doses that elicit priming are moderate in that blood alcohol concentrations only have to peak in a range of .03 g% to .06 g% (levels well below the level of legal intoxication). Not surprisingly, individual differences in priming effects are often observed, with greater priming effects associated with greater likelihood to abuse alcohol (Fillmore, 2001). At least in animal research, priming effects may be attributable to alcohol-induced activation of neural reward mechanisms (Robinson and Berridge, 1993). The priming dose may also decrease the effectiveness of inhibitory control mechanisms (Fillmore, 2003).

Past studies of alcohol-induced priming of alcohol consumption often focused on the role that priming plays in alcohol dependence. The effect that priming plays on binge drinking is less well understood, and several unanswered questions remain. In particular, past studies did not take into account three factors: 1) whether "desire for drug" ratings actually coincided with the real-world consumption of the drug, 2) whether binge drinkers are more prone to priming effects compared with more moderate drinkers, and 3) how significant a role the disinhibition by the initial dose of alcohol plays in the subsequent consumption of alcohol. A recent investigation that incorporated all three of these factors has been reported.

Weafer and Fillmore (2008) tested the hypothesis that social drinking college students who experience greater impairment in inhibitory control from a dose of alcohol were also more likely to consume alcohol when given access to the drug. In this study, participants included 13 binge and 13 non-binge social drinking college students. Participants were brought into the laboratory for three different sessions. On two of the sessions, all of the individuals were tested on a cued go/no-go computerized task, which measures inhibitory control, as described in greater detail in Chapter 11. Participants completed the computerized task following a 0.65 g/kg dose of alcohol on one of the sessions and following a placebo dose on the other session. Finally, on the third session, participants engaged in a taste-rating task which is an unobtrusive laboratory measure for determining ad libitum alcohol consumption (Marlatt et al., 1973; Marczinski et al., 2005). Participants were told that they would participate in a beer taste-rating task, the purpose of which was to determine the preferences of college students for common beers. The participants then had 90 minutes to sample several different beers and complete a questionnaire about their thoughts on the qualities of the beverages (e.g., smell, taste, carbonation etc.). The purpose of this session was to determine: 1) how much does the participant consume, and 2) what blood alcohol concentration does the person achieve. Previous work has shown that the amount typically consumed and the blood alcohol concentration achieved during an ad libitum drinking session strongly correlates with self-reported drinking habits (Marczinski et al., 2005).

The results of the Weafer and Fillmore (2008) study indicated that the individuals who displayed greater impairment of inhibitory functioning in response to the alcohol also consumed greater amounts of alcohol when given ad libitum access to it during the beer taste-rating task. This relationship between alcohol-induced impairment of inhibitory control and ad libitum drinking supports the hypothesis that greater reduction in inhibitory control from alcohol could seriously compromise the social drinker's ability to stop the ongoing self-administration of alcohol. Thus, drinking may continue unabated and a binge results. While these results are preliminary, it does suggest that there may be a strong relationship between inhibitory control and alcohol consumption. When inhibitory control is compromised, loss of control over drinking may be the result. More importantly, this pattern is not limited to severe alcoholics. The same situation also appears to apply to social drinking college students, albeit to a lesser severity than observed among those with serious alcohol dependence.

INHIBITORY CONTROL AND THE ANTERIOR CINGULATE

Thus far, links between the alcohol-induced impairment of inhibitory control mechanisms in the brain and various behaviors have been made. This chapter highlighted some of the research findings that have made those links with behaviors such as poor decision-making in the context of impaired driving, aggression, sexual behavior and the further consumption of alcohol. However, the behaviors described in this chapter may only be the tip of the iceberg. Failures of self control are not limited to those described herein. Societal problems such as drug abuse, crime, and obesity are only a few of the problems which can occur when an individual fails to control behavior, thus revealing how humans are uniquely defined by this self-control trait and yet often lack this crucial ability (West, 2007).

Why is maintaining self control such a difficult thing to do, and what region of the brain is responsible for this task? One recent study provides insights regarding this question. It has been suggested that we have limited brain resources to control ourselves and all acts of control draw from these same resources. Thus, if we are drawing heavily on these resources for use in one domain, another domain may suffer. For example, if one is studying hard for an exam, his plans to keep to a diet may fail. Inzlicht and Gutsell (2007) examined this idea that self control is a limited resource, and investigated the role of a part of the brain's frontal lobe called the anterior cingulate cortex in self control using electroencephalography (EEG). EEG involves measuring the brain's electrical activity using electrodes attached to the scalp. In this experiment, participants completed two phases. First, the researchers had participants 'deplete their self control source' by asking participants to suppress their emotions while watching an upsetting movie. Participants reported their success in suppressing their feeling on a scale from one to nine. Second, participants completed a Stroop task, which involves naming the color of printed words (i.e., saying 'green' when reading the word 'red' which is printed in green type).

The researchers observed that participants who successfully suppressed their emotions during the movie performed worse on the Stroop task. It is possible that the participants who used up their resources for self-control while holding back their tears during the film had little self-control left for the Stroop task. The EEG results from the Stroop task echoed the behavioral results. Weaker activity in the anterior cingulate of the brain was observed for those participants who had successfully suppressed their feelings. Normally, increased brain activity in the anterior cingulate is observed with Stroop task performance. Thus, after engaging in one act of self-control, this brain system seemed to fail during the next act (Inzlicht and Gutsell, 2007). It is not surprising that alcohol impairs self-control as this brain function appears to be limited in its ability, even when individuals are in a sober state.

CONCLUSION

Decreased inhibitory control may have widespread effects on other behaviors. Alcohol-induced disinhibition may result in failures of self-control such as aggressive behavior, inappropriate sexual behavior, and further drinking. A vicious cycle may be at play between disinhibition and alcohol consumption contributing to binge drinking, with only a small amount of alcohol being required to decrease inhibitory control and lead to further drinking. Although further research is needed, it appears that one region of the brain, the anterior cingulate cortex, may be necessary for maintaining inhibitory control. However, the sensitivity of this region to depleting effects of other situations requiring inhibitory control may provide some insight as to why self control is so vulnerable to the effects of alcohol.

Chapter 13

BINGE DRINKING AND ALCOHOL TOLERANCE

The repeated administration of any given dose of a drug, including alcohol, often results in a reduced response to that drug, a phenomenon called tolerance (Maisto et al., 2008). The Diagnostic and Statistical Manual of Mental Disorders (DSM-IV) defines tolerance as either the need for markedly increased amounts of alcohol to achieve intoxication or the desired effect or markedly diminished effects with continued use of the same amount of alcohol (American Psychiatric Association, 1994, 2000). As tolerance develops, higher doses of alcohol might be needed to reinstate the initial effect, which is why alcohol tolerance has become recognized as a factor that may contribute to alcohol abuse and dependence by encouraging the use of escalating doses (American Psychiatric Association, 1994). Tolerance is obviously a concern with chronic heavy drinkers. By contrast, binge drinkers generally don't drink daily. However, binge drinkers do drink large amounts of alcohol with a certain amount of regularity. Thus, researchers have asked whether the development of tolerance may be occurring over time in these binge-drinking individuals.

There are three distinct types of alcohol tolerance. *Dispositional tolerance* refers to an increase in the rate of metabolism of alcohol as a result of its regular use. Dispositional tolerance may be mostly evident in chronic heavy drinkers as alcohol metabolism becomes slightly more efficient with the daily ingestion of doses of alcohol. *Functional tolerance* (also called pharmacodynamic tolerance) refers to the decreased behavioral effects of alcohol as a result of its regular use. With functional tolerance, the brain and other parts of the central nervous system become less sensitive to the effects of alcohol. Functional tolerance is further subdivided into acute and protracted subtypes. Acute tolerance is measured within the course of action of a single dose of alcohol while protracted tolerance refers to the effects of a given dose of alcohol when it is administered chronically. Functional tolerance may be developing in both chronic and binge drinkers. Finally, *behavioral tolerance* (also called learned tolerance) involves a behavioral adjustment made by an individual in response to the alcohol (Maisto et al., 2008). For example, if an individual learns that alcohol intoxication makes him slur his speech, he may speak more slowly when drinking to compensate for this effect of alcohol. Even lighter drinkers may develop behavioral tolerance.

Tolerance is an interesting and somewhat complicated topic which is why it is still unclear to what degree binge drinkers develop any or all of the above three types of tolerance. It is well understood that alcoholics have developed significant tolerance to alcohol and this tolerance is part of the addiction. However, the nature of binge drinking (intermittent high levels of alcohol consumption followed by periods of no drinking) is unlike the alcohol consumption patterns of individuals who are dependent on alcohol and drink daily. If binge drinking leads to appreciable development of tolerance, then binge drinking may lead to further risk for dependence on alcohol. Research on alcohol tolerance is complicated by the fact that tolerance may develop to some effects of a drug but not others.

WHY DOES TOLERANCE DEVELOP?

No one theory fully accounts for the development of tolerance. However, some theories account for portions of findings of the development of tolerance. The current thinking is that tolerance development is multifaceted and involves both learning and biological processes (Maisto et al., 2008). Biological processes are implicated in the "adaptation-homeostasis hypothesis" which assumes that any drug, including alcohol, acts on specific cells in the central nervous system (CNS). Since plasticity is a feature of the CNS, the cells become adapted to the presence of alcohol with repeated exposure to it. Adaptation allows cells to maintain normal functioning when alcohol is present. The reason tolerance is observed is because more alcohol is needed to disrupt cell functioning (Cicero, 1980).

Learning processes are also implicated in the "drug compensatory reactions" explanation of tolerance. For example, an individual who takes a drug repeatedly in one location (Person A) may develop greater tolerance than another individual who takes the same kind and amount of drug repeatedly, but in many different locations (Person B). Recall that the adaptation-homeostasis hypothesis would not predict any differences in tolerance in these two people. Findings from both human and animal studies have shown that Person A usually has much greater tolerance than Person B, which provides evidence that learning is part of the development of tolerance. Cues associated with drug administration (such as the sight of the local bartender or the clinking of glasses in the local bar) become associated with alcohol administration and the body works to counteract the effect of the alcohol that will soon enter the system (Hinson, 1985; Siegel, 1989).

ACUTE TOLERANCE

Although tolerance to alcohol usually develops over time and over several drinking sessions, it can also be observed within a single drinking session. This is known as acute tolerance (Tabakoff et al., 1986). Upon completion of drinking, the amount of alcohol in an individual's body (measured as the amount of alcohol in the blood) rises to some peak level (Maisto et al., 2008). Following this peak, blood alcohol level declines at a slower rate than the initial rise. Blood alcohol declines more slowly than it rises because the liver can only metabolize alcohol at a constant rate of .015 g% per hour. This metabolic rate is roughly equivalent to one standard drink per hour. The metabolic rate is independent of the amount of alcohol consumed, the person's body weight, or the person's metabolic needs. However, individuals who are extremely heavy drinkers may have a slightly faster metabolic rate, but an extended drinking career is usually needed to observe this faster metabolism (i.e., dispositional tolerance). The constant metabolic rate of alcohol driving the decline in BAC differs from the fact that the rise in BAC for a given amount of alcohol that is consumed does depend on body weight (Kinney, 2009; Maisto et al., 2008). Typically, acute tolerance is said to be evident when the effects of alcohol are greater when measured soon after alcohol administration than when measured later, even if the blood alcohol concentration (BAC) is the same at both times (Beirness and Vogel-Sprott, 1984; Bennett et al., 1993; Hiltunen and Jarbe, 1990).

Figure 13-1 plots the breathalyzer readings of individuals who participated in an acute tolerance alcohol dose administration study (Marczinski and Fillmore, in press). Breathalyzer readings are used in alcohol research as they are a reliable indicator of blood alcohol levels. This is because removal of alcohol from the body begins as soon as the alcohol is absorbed into the bloodstream. Small amounts of alcohol leave the body unmetabolized through breath, sweat, and urine. The proportion of alcohol that is exhaled into the air has a constant and predictable relationship to the blood alcohol concentration. As such, one can measure breath alcohol using a machine called a breathalyzer and have a good indication of the actual blood alcohol level (Kinney, 1990). Breathalyzers are used by police when stopping drivers for possible impaired driving. They are also used by alcohol researchers when testing for acute tolerance. In this acute tolerance study, alcohol was administered within the first five minutes for participants who had not eaten for four hours. Blood alcohol rose swiftly and declined more slowly, as is typically the case. To determine whether acute tolerance was present, participants' performance was measured on two tests. The first test was on the ascending limb of the blood alcohol curve and the second test was on the descending limb of the blood alcohol curve, when BACs were comparable to the first test (Marczinski and Fillmore, under review).

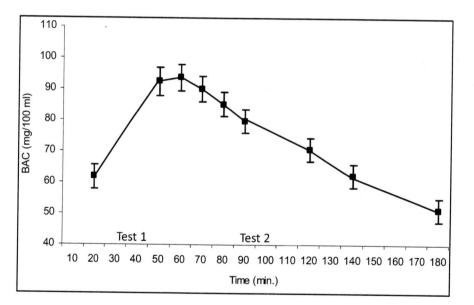

Figure 13.1. Mean blood alcohol concentrations (BACs) under alcohol at each interval when breath samples were obtained. The ascending limb test was begun at 30 min. following alcohol dose administration and the descending limb test was begun at 90 min. following alcohol administration. Tests 1 and 2 were at comparable mean BACs. Time recording began upon the first sip of alcohol. Participants were instructed to drink the alcoholic beverage within five minutes. Vertical bars show standard errors of the mean (Marczinski and Fillmore, in press).

One measure obtained was a rating of willingness to drive. Participants stated that they were more willing to drive an actual vehicle on the descending limb test, compared to the ascending limb test, despite equivalent BACs. This change in subjective reaction to alcohol across the two limbs of the blood alcohol curve indicates the development of acute tolerance.

The development and magnitude of observed acute tolerance appears to be an important feature in a social drinker indicating who might develop more serious problems with alcohol. For example, Beirness and Vogel-Sprott (1984) found that the greater the extent of acute tolerance, the greater the development of overall tolerance in the drinker. Thus, an individual who appears significantly less impaired by alcohol on the descending limb of the blood alcohol curve, compared to the ascending limb test, is also likely to be quite tolerant to alcohol in general. Since tolerance to alcohol is a feature that puts an individual at risk for alcohol abuse and dependence (American Psychiatric Association, 1994), observations of acute tolerance are a hallmark feature that a drinker may develop alcohol abuse and dependence.

ACUTE TOLERANCE DEVELOPMENT IN BINGE AND SOCIAL DRINKERS

The findings of Beirness and Vogel-Sprott (1984) that acute tolerance and tolerance in general may be evident in social drinkers have led other investigators to investigate this issue. Previously, it was thought that tolerance was the concern of the chronic alcoholic. However, this appears not to be the case, and may explain some of the risky and dangerous behaviors exhibited by high school and college binge drinkers. For social drinkers, several studies have reported acute tolerance to perceived intoxication (i.e., feeling drunk). For example, several studies have reported that social drinkers report feeling less intoxicated and less stimulated by alcohol on the descending limb of the blood alcohol curve compared with the ascending limb, despite equivalent BACs (Evans and Levin, 2004; Fillmore et al., 2005; Hiltunen, 1997; King and Byars, 2004; King et al., 2002; Martin et al., 1993; Portans et al., 1989; Rose and Grunsell, 2008). Acute tolerance to the subjective effects of alcohol appears also to be related to drinking habits. As typical alcohol use increases, acute tolerance to perceived intoxication develops more rapidly (Hiltunen, 1997; King and Byars, 2004). Evans and Levin (2004) found that moderate drinkers were less likely to experience negative subjective alcohol effects (e.g., feeling tired and sleepy) on the descending portion of the blood alcohol curve compared to their lighter drinking peers. Rose and Grunsell (2008) also recently reported that binge drinkers were more tolerant to the sedative and lightheaded effects of alcohol, relative to non-binge drinkers.

Despite these robust findings, little connection has been made between the development of acute tolerance to the interoceptive cues of intoxication in binge drinkers and their subsequent risky decisions, such as whether to drive. While decisions to drive after drinking are likely based on a number of personal, social, and environmental factors, one factor that has not been fully considered is the development of acute tolerance to the drinker's self-perception of intoxication. Since the majority of arrests for driving under the influence occur late at night (Shore et al., 1988), decisions about whether to drive under the influence are likely to occur on the descending portion of the blood alcohol curve, which may be the time point where misperceptions of level of intoxication are magnified in binge drinkers.

In a recent study, the degree to which alcohol impaired binge and non-binge drinkers' driving performance was examined in comparison with their own self-appraisal of their level of intoxication and fitness to drive. It was predicted that binge drinkers would demonstrate greater acute tolerance to the perceived intoxicating effects of alcohol compared to non-binge drinkers. Furthermore, it was predicted that this interoceptive cue of level of intoxication might contribute to the drinkers' decision on how willing they are to drive, accounting for the heightened impaired driving rates in binge drinkers (Marczinski and Fillmore, in press).

Both binge (N = 18) and non-binge (N = 10) social drinking college students were brought into the laboratory to participate in a study of acute tolerance to alcohol on driving performance (using a driving simulator) and subjective ratings of willingness to drive. Participants attended two different sessions. In one session the participant received a dose of alcohol (0.65 g/kg) and in another session the participant received a placebo (a drink that looked and smelled like alcohol, but contained no alcohol). Binge and non-binge drinkers were classified using the well-established Wechsler definition of binge drinking, with binge drinking defined as the consumption of five or more drinks per episode for men and four or more drinks per episode for women (Wechsler et al., 1994, 1998, 2000). This measure assumed that drinking is to intoxication if the individual meets or exceeds the five/four drinks criterion. All participants were tested twice during each session: at 30 minutes and at 90 minutes following the onset of alcohol administration. Based on previous work that had estimated the rise and fall in blood alcohol over time (Fillmore et al., 2005), participants were expected to achieve an average BAC of .065 g% at 30 minutes that would continue to rise to an approximate peak of .08 g% at 60 minutes and descend back to .065 g% by 90 minutes. To ensure that this actually occurred, BAC was measured approximately every 10 minutes during the alcohol session, as shown in Figure 13-1. Breath samples were also obtained at the same times during the placebo session, ostensibly to measure participants' BACs. Participants completed the subjective effects measures either before or after the driving test, with order of task administration counterbalanced between participants.

The driving simulator provided a realistic experience of driving for the participants. The participants completed a 20-minute driving course which covered approximately 19 miles and included a winding road, hills and buildings by the side of the road. Participants were also asked to rate their subjective reactions to the dose (alcohol or placebo) that was administered. On both tests, participants were asked to rate how intoxicated they felt and how willing they were to drive an actual car at the time of the rating. The subjective intoxication scale was used to test participants' perceived level of intoxication (Fillmore and Vogel-Sprott, 2000). This scale uses units familiar to all participants by asking participants to rate their perceived level of intoxication based on how many drinks that they thought they had consumed relative to a bottle of beer containing 5% alcohol. The scale ranges from 0 to 10 bottles of beer, in 0.5 bottle increments. The willingness to drive rating is a visual analogue scale that asks participants to rate how willing they are to drive an actual vehicle at the time they are completing the questionnaire (Beirness, 1987). Participants indicate their response ranging from 'not at all' to 'very much' by placing a vertical mark through a 100-mm line.

The results obtained were consistent with previous findings of acute tolerance to the subjective effects of alcohol in heavy social drinkers. Compared with placebo, alcohol increased ratings of intoxication and decreased willingness to drive in both binge and non-binge drinkers. However, during the descending phase of the blood alcohol curve, the binge drinkers showed acute tolerance to alcohol's effects on subjective intoxication. Thus, binge drinkers felt less drunk on the descending limb alcohol test, compared to the ascending limb test, despite equivalent BACs. This effect was accompanied by an increased rating of willingness to drive on the descending limb test, compared to the ascending limb test. Figure 13-2 illustrates the willingness to drive ratings for both the binge and non-binge drinkers. On the first alcohol test (during the ascending portion of the blood alcohol curve), both binge and non-binge drinkers were unwilling to drive (they gave low willingness to drive ratings). However, binge and non-binge drinkers diverged in the willingness to drive ratings by the second alcohol test. Despite equivalent BACs on both tests, only the binge drinkers became more willing to drive on the descending limb of the blood alcohol curve.

Of course, the results on the willingness to drive ratings leave one wondering whether the development of acute tolerance to the subjective ratings also coincided with actual changes in driving behavior. If acute tolerance developed in one domain (subjective feelings of intoxication), perhaps acute tolerance also developed in another domain (actual driving skill). However, acute tolerance in driving skill was not observed. Instead, driving performance was impaired by alcohol on both limbs of the blood alcohol curve. Figure 13-3 illustrates one measure from the simulated driving test, mean within-lane deviation (a measure of swerving). Alcohol impaired driving (i.e., more swerving was observed) under both tests. In fact, driving performance was actually worse under the second descending limb test compared to the first ascending limb test under alcohol, despite equivalent BACs. Other aspects of driving were similarly impaired by alcohol and no evidence of acute tolerance was observed for any driving measure in any of the participants.

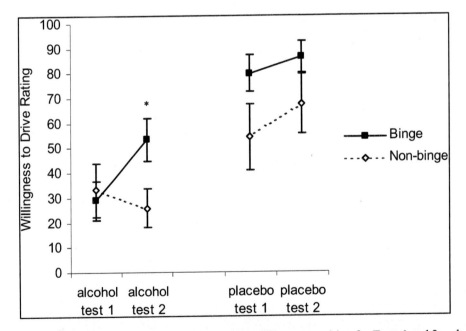

Figure 13.2. Mean self-reported subjective responses of willingness to drive for Tests 1 and 2 under the 0.0 g/kg (placebo) and 0.65 g/kg alcohol dose conditions for the binge and non-binge drinkers. Vertical bars show standard errors of the mean (Marczinski and Fillmore, in press). * Significant difference between binge and non-binge drinkers ($p < .05$).

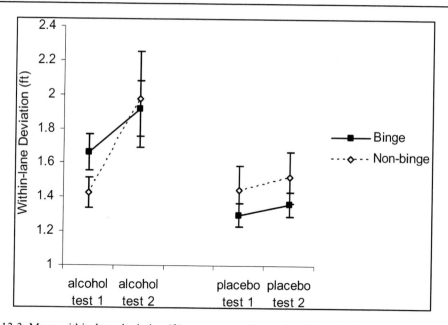

Figure 13.3. Mean within-lane deviation (ft), a measure of swerving, from the simulated driving task for Tests 1 and 2 under the 0.0 g/kg (placebo) and 0.65 g/kg alcohol dose conditions for the binge and non-binge drinkers. Vertical bars show standard errors of the mean (Marczinski and Fillmore, in press).

Thus, decreased perceived level of intoxication may lead to a greater likelihood of driving among binge drinkers, possibly accounting for the greater incidence of impaired driving in this demographic group. Changes in perceived intoxication are not accompanied by changes in driving skill, which still remains extremely poor under alcohol for binge and non-binge drinkers alike.

Differential development of acute tolerance to the subjective effects of alcohol in binge and non-binge drinkers is not a surprising finding considering typical drinking patterns in these two groups of participants. In this study, all participants were asked about their typical drinking patterns and also asked to record their recent drinking activities using a calendar. The binge drinkers reported that they typically drank more than twice the number of drinks than the non-binge drinkers on an average drinking occasion. The binge drinkers also reported routinely drinking to intoxication while the non-binge drinkers did not. Since the relationship between alcohol consumption patterns and the subjective effects of alcohol have been reported before, the present findings were consistent with recent reports that binge drinkers feel less impaired by alcohol compared to more moderate drinkers (Brumback et al., 2007; Holdstock et al., Marczinski et al., 2007, 2008; Rose and Grunsell, 2008).

Furthermore, the evidence that binge drinkers' acute tolerance to the subjective effects of alcohol was not accompanied by tolerance to their actual behavioral impairment has important implications for understanding why this group is at-risk for driving-related offenses. Based on times of arrest for impaired driving (Shore et al., 1988), most decisions to drive will be made on the descending limb of the blood alcohol curve when binge drinkers feel less intoxicated. Interoceptive cues concerning one's level of intoxication likely plays a role in decisions to drive. This might be less of a problem for non-binge drinkers who have not developed much appreciable acute tolerance to either the subjective effects of alcohol or the impairing effects of alcohol on driving performance. However, the failure to fully appraise one's level of intoxication could be a significant problem for binge drinkers, which might be particularly difficult during the descending phase of the BAC curve.

Why would acute tolerance develop for one domain (i.e., subjective intoxication) and not another (i.e., driving performance)? Schweizer and Vogel-Sprott (2008) reviewed the available studies on acute tolerance and recovery of cognitive performance in social drinkers. They found that in many studies of the effects of a dose of alcohol, basic motor skills (e.g., using a stylus to track a moving target) can recover from impairment as BAC declines, indicating the development of acute tolerance. However, cognitive performance appears to function differently. Speed of cognitive performance usually recovers from impairment to drug-free levels during the decline in BAC. However, alcohol-increased errors fail to diminish, and in fact may increase in some cases. Thus, speed of cognitive processing tends to develop acute tolerance, but no such tendency is observed in accuracy. Schweizer and Vogel-Sprott (2008) suggested that differential development of acute tolerance may reflect the fact that some areas of the brain are more sensitive to the impairing effects of alcohol and take longer to recover from its effects. Since driving performance is a complicated task requiring a variety of motor and cognitive skills, it is not surprising that acute tolerance was not observed for the alcohol-induced impairment of driving performance.

There may be safety risks when acute tolerance develops to some aspects of alcohol-induced impairment (e.g., subjective intoxication, basic motor skills, reaction time) but not others (e.g., accuracy). Self-assessment of functioning under alcohol provide the basis for decisions about engaging in potentially hazardous activities (e.g., walking home in the dark alone, driving, and jumping from a height). If all skills are impaired, judgment may be better. However, if the person feels less intoxicated and their reaction time has recovered (so that they don't feel slow and sluggish), they may erroneously believe that they are capable to doing a variety of tasks that they shouldn't based on their intoxication level.

SENSITIZATION

Sensitization refers to the opposite process from tolerance, which is why it is also sometimes called "reverse tolerance". Tolerance refers to the diminished effect with repeated use of alcohol or increased amounts of alcohol needed to achieve intoxication. By contrast, with repeated use of a drug, some neural pathways may heighten (sensitize) the reward value of the drug (Robinson and Berridge, 2003). There has been much less investigation of sensitization and the phenomenon may be less common than tolerance. However, some preliminary evidence may indicate that some sensitization may be occurring in binge drinkers. As discussed in chapter 11 on behavioral control, Marczinski et al. (2007) found that binge drinkers were more disinhibited (i.e., they acted more impulsively) by an acute dose of alcohol compared to their more moderate drinking peers. The binge drinkers also reported feeling more stimulated by the alcohol compared to the non-binge drinkers. There were no differences between binge and non-binge drinkers on either disinhibition or stimulation ratings following the administration of a placebo. An intriguing aspect of this finding of a heightened disinhibitory reaction to alcohol in binge drinkers is that the finding runs counter to what would be predicted based on basic principles of pharmacological tolerance, and therefore may be evidence of sensitization. The binge drinkers in the Marczinski et al. (2007)

study reported higher levels of alcohol consumption per episode and drank more frequently. From a pharmacological perspective, these individuals would be expected to be more tolerant to the acute effects of the alcohol dose since their body should be more familiar with that amount of alcohol. However, the binge drinkers actually displayed more impairment in inhibitory control and greater increases in stimulation following alcohol consumption, consistent with the phenomenon of sensitization (see also King et al., 2002). There is still relatively little that is known about the actual behavioral tolerance and sensitization that develops in binge drinkers and whether it is uniform across various domains of behavioral functioning. Perhaps binge drinking does not afford any appreciable tolerance to inhibitory mechanisms that control behavior and instead leads to sensitization. As stated earlier, the understanding of when tolerance and sensitization occur is an issue that researchers are still working to understand.

CAN TOLERANCE AND SENSITIZATION BE REVERSED?

Once tolerance or sensitization to alcohol has become established, it is not irreversible. We have heard many anecdotal reports from students who state that they developed great tolerance to alcohol over a certain period of time (e.g., during school) but then lost that tolerance when they stopped their heavy drinking (e.g., while working over the summer). Similarly, many college graduates make claims that 'they no longer have the tolerance that they once had as a student'. A period of abstaining from any drug, including alcohol, increases the user's sensitivity to drug effects that he or she may have become highly tolerant to in the past. In general, acute tolerance reverses in a short period of time whereas protracted tolerance requires a longer and more extended period of abstinence to reverse (Maisto et al., 2008). However, if the individual resumes the alcohol consumption pattern of the past, reacquisition of tolerance often occurs more quickly than when it developed the first time around (Kalant et al., 1971). Interestingly, behavioral (learned) tolerance is the most difficult to reverse since the behavior would have be unlearned (Maisto et al., 2008).

CONCLUSION

As typical drinking patterns increase, tolerance to the effects of alcohol occurs. There are varieties of tolerance, such as acute tolerance, which has been widely studied in social drinkers. When blood alcohol rises and falls, acute tolerance is said to have developed when impairment is less on the declining limb of the blood alcohol curve compared to the ascending limb, despite equivalent BACs. Acute tolerance is more pronounced in binge drinkers compared to non-binge drinkers. Greater acute tolerance is associated with the risk for alcohol abuse and dependence. Acute tolerance does not appear to develop to all aspects of alcohol-induced impairment. Robust acute tolerance has been reported for the subjective effects of alcohol (e.g., feeling less intoxicated on the declining limb of the blood alcohol curve compared the ascending limb, despite equivalent BACs). However, acute tolerance has not been reported for other cognitive skills such as driving. As such, the development of acute tolerance sets up a significant safety risk for binge drinkers. On the declining limb of the blood alcohol curve, binge drinkers may feel less intoxicated and think that they are capable of driving home. Driving skills would nevertheless remain impaired. These findings may account for the heightened impaired driving rates in binge drinkers.

Chapter 14

BEFORE THE FIRST DRINK: AT RISK INDIVIDUALS FOR THE BINGE CONSUMPTION OF ALCOHOL

Identification of the precursors to initiation of underage alcohol consumption and binge drinking would be extremely helpful in targeting prevention programs towards young people who might start engaging in these health-risk behaviors. There are several factors covered in this chapter that are strong predictors of binge alcohol consumption in young people. A risk factor is any characteristic that increases the probability of a particular outcome. Thus, several variables have been identified that increase the probability that an individual may become a binge drinker. Such factors include personality traits, such as impulsivity, as well as other factors such as family dynamics. While widely varying in their form, identification of these factors might prove helpful in identifying those children who are most likely to develop a binge drinking problem and/or a later substance abuse problem. Of course no single factor in and of itself is a perfectly reliable causal predictor of future binge alcohol consumption. Many college students who develop binge drinking problems exhibit none of the below described characteristics, which highlights the fact that there are significant social factors at play at the time binge drinking is occurring. Nonetheless, several important risk factors of binge drinking behavior have been identified. Those risk factors are detailed below.

IMPULSIVITY

One strong line of research on the relationship between personality traits and alcohol use and abuse is the link between impulsivity and alcohol consumption and alcohol abuse (Rubio et al., 2008; von Diemen et al., 2008). Personality refers to a cluster of characteristics that describe the ways in which an individual typically thinks, perceives, feels, and acts. While there are variations in how a person acts in different situations, these personality characteristics are thought to be fairly constant. Impulsivity is a personality trait characterized by acting on impulse, nonplanning, responding with a lack of reflectiveness, liveliness, and risk taking (Acton, 2003; Magid et al., 2007). All young children are impulsive by nature. A 2 year-old child will grab a toy from another child, without waiting his or her turn. This impulsive action, normal for a young child, is inappropriate as the child gets older. Control over behavior and delay of gratification occur with development and coincide with the growth of the frontal lobes of the brain, particularly the prefrontal cortex (the very front of the brain that is presumed to control behavior) (Franken et al., 2008). Adults who have poor activity in the prefrontal cortex or adults who have brain damage in this region are often impulsive and have poor control over their thoughts and actions (Cato et al., 2004; Franken et al., 2008). Findings from experimental, cross-sectional, and longitudinal research point to the importance of impulsivity as a vulnerability factor for substance use and abuse (Acton, 2003). Impulsivity also appears to be an important personality factor associated with binge drinking (Ryb et al., 2006).

What is behavioral control and what is different in the brain of someone who has poor impulse control? Various theoretical models have portrayed behavioral control as being governed by two processes in the brain. These two processes are thought to be distinct and operate fairly independently. One process is thought to activate behavior and one is thought to inhibit behavior (Fowles, 1987; Gray, 1976, 1977; Logan and Cowan, 1984; Patterson and Newman, 1993; Quay, 1977). The two processes act in opposition to one another and the relative strength of each is assumed to determine behavioral control. Thus, if the activation process dominates, a behavior will be observed. If the inhibition process dominates, a behavior will be suppressed. When a child grabs a cookie off the table, activation is dominating. If the child withholds their response despite wanting a cookie, inhibition is dominating. With this model, inhibition is an active cognitive process that demands resources. Thus, not acting in some cases can be quite difficult, especially if the reward is very appealing. When one observes overactive, impulsive behavior in an individual, it is assumed that behavioral inhibition is deficient (Logan et al., 1984). Deficient behavioral inhibition is considered to be a primary mechanism by which an acute dose of alcohol impairs self-control (Fillmore, 2003; Pernanen, 1993). It has been argued that disinhibited individuals often have a strong reward focus (McCarthy et al., 2001). For example, the person that is disinhibited may have another drink because it tastes good and it makes him feel good, even though he needs to get home to study for an exam the next day. For the disinhibited individual, the immediate reward of the drink is much more compelling than the more delayed reward of getting a good grade on an exam.

Impulsivity and disinhibition are somewhat intermingled concepts and the term that is chosen often depends on who is using it and in what context. Impulsivity is often described by researchers as a personality trait and is thought to be an aspect of a person that remains relatively constant over time and in various environments. Thus, some individuals are, by nature of their personality, more impulsive than others. By contrast, inhibitory control and disinhibition are often described by cognitive neuroscientists as aspects of brain function that can be changed by drugs, or as cognitive processes that become less effective in certain contexts. Most drugs of abuse, including alcohol, have been shown to impair inhibitory control (Fillmore, 2003). An acute dose of alcohol will temporarily impair inhibitory functioning in all social drinkers (Marczinski and Fillmore, 2003a, 2003b, 2005). Evidence for impulsivity and disinhibition include failures of self-control. This can be assessed in multiple ways including giving participants questionnaires or by tasks given to participants to complete in the laboratory.

There are several common and widely-used questionnaires that assess the personality trait of impulsivity. One common questionnaire is the Eysenck Impulsiveness Questionnaire. Eysenck proposed the highly influential biologically-based model of personality. He conceptualized personality as three biologically-based traits of temperament: extraversion, neuroticism and psychoticism. Impulsivity was thought to be part of psychoticism and

Eysenck suggested that testosterone levels underlay this trait, with higher levels of psychoticism associated with higher levels of testosterone (Eysenck and Eysenck, 1985). The Eysenck Impulsiveness Questionnaire assesses the personality trait of impulsivity by posing 19 questions to which the participant answers yes or no, depending on whether that statement applies to them (Eysenck et al., 1985). For example, the participant would read the statement, "Do you often do things on the spur of the moment?" and would respond either yes or no. The higher the number of affirmative responses, the higher the level of self-reported impulsivity.

Another questionnaire used to assess impulsivity is the Barratt Impulsivity Scale-11 (BIS-11). This questionnaire asks participants to rate how typical 30 different statements are for them. Participants read each statement and then respond on a 4-point Likert scale ranging from Rarely/Never to Almost Always/Always (Patton et al., 1995). For example, the participant would be asked to respond on how typical the statement, "I make up my mind quickly" is for them. The higher the summed scores for all of the items, the higher the level of self-reported impulsiveness. In addition to the total score, there are six factors than can be obtained from the questionnaire that assess different aspects of impulsivity, including Attention (focusing on the task at hand), Motor Impulsiveness (acting on the spur of the moment), Self-Control (planning and thinking carefully), Perseverance (maintaining a consistent lifestyle), and Cognitive Instability (thought insertions and racing thoughts). Marczinski et al. (2007) compared binge and non-binge drinking college students on their responses to the BIS-11 questionnaire. The results indicated that binge drinkers had higher BIS-11 scores compared to their more moderate drinking peers. Thus, binge drinkers reported that they were more impulsive. Furthermore, binge drinkers differed from the non-binge drinkers on the factors of Attention, Motor Impulsiveness and Perseverance. Binge drinkers reported that they had more trouble focusing on tasks at hand, were more likely to act on the spur of the moment and were less likely to maintain a consistent lifestyle, such as living in the same location (Marczinski et al., 2007).

While there is only a small amount of evidence that impulsivity is associated with binge drinking in young adults, there is a larger amount of evidence that impulsivity is associated with alcoholism. It has been reported that adult individuals who score high on measures of impulsivity are more likely to have an alcohol use disorder (Trull et al., 2004). This finding is not surprising considering that one of the core features of an alcohol use disorder is loss of control over drinking. Thus, an individual who acts without consideration of the consequence may continue to drink alcohol, despite the risk of being fired at work, losing their spouse, or being arrested for impaired driving. However, the cause and effect status of this relationship between impulsivity and alcohol use disorders is unclear. Are impulsive people at greater risk of developing an alcohol use disorder or does alcohol use trigger impaired control? Since both directions of causation have been shown, this relationship could even be circular in nature. Thus, an impulsive person (with poor development in the frontal lobes) may start drinking heavily. The alcohol consumption may lead to further damage to the frontal lobes which results in even greater impulsivity leading to out of control drinking.

To examine whether the personality trait of impulsivity is the precursor to later alcohol use problems, a prospective risk study is helpful. One such prospective study was completed by Cloninger et al. (1998) in Sweden. The authors recruited 431 children at age 11 and gave them a detailed behavioral assessment. Then, the authors reevaluated all of these individuals at age 27, looking for evidence of alcoholism or an alcohol abuse problem. They found that childhood impulsivity levels at age 11 were a major risk factor for adult alcohol abuse.

Furthermore, Jones (1968) found that impulsivity correlated highly with problem drinking later in life. McCarthy et al. (2001) also found that trait levels of disinhibition related to self-reported drinking.

Are there gender differences in the role that impulsivity plays in future alcohol abuse? Petry et al. (2002) examined this issue. It is already known that alcohol dependence has a clear familial component. Having a parent that is an alcoholic puts one at increased risk of alcoholism. Petry et al. (2002) compared impulsivity in individuals who differed in respect to paternal history of alcohol dependence. By comparing individuals who did or did not have a father who was an alcoholic and also examining impulsivity of the participant, the authors were hoping to provide evidence of familial vulnerability to impulsivity. All participants completed a delayed discounting task, which is a laboratory measure of impulsivity. In this task, participants are offered choices between monetary rewards (e.g., \$34) available immediately and larger rewards (e.g., \$50) available after various delays ranging from one week to six months. In this task, more impulsive individuals tend to choose immediate rewards over the delayed rewards, even though the delayed rewards are often more money. The authors did find that women with a positive familial risk of developing an alcohol use disorder (i.e., their genetic risk of developing alcoholism was high) were also more impulsive on the delayed discounting task relative to those with a negative family history of alcoholism. However, the authors did not find this difference with their male participants. Why impulsivity was higher in females with a genetic risk of alcoholism compared with females without a positive family history, while this pattern was not seen with males, is not clear. Future research will need to determine if the gender difference is replicable across other studies and what this gender difference means for understanding the role that impulsivity plays in the abuse of alcohol.

One extreme form of impulsivity is attention deficit hyperactivity disorder (ADHD). A diagnosis of ADHD in a child may be made if behavior is characterized by heightened impulsivity and impaired inhibitory and attentional mechanisms (Krain and Castellanos, 2006). These cognitive and behavioral problems often adversely impact a child's performance in school as evidenced by poor grades and behavior in the classroom that is disruptive, such as an inability to sit still or pay attention (Frazier et al., 2007). Most diagnoses of ADHD occur in childhood, with recent studies suggesting that the disorder persists in adulthood in approximately 60 to 80% of cases (Barkley et al., 2002; Weiss et al., 1999; Wender, 1995). In adults with ADHD, the cognitive impairments associated with the disorder are problematic in the workplace and for overall career success. Adults with ADHD often have considerable difficulty in organizing job-related activities, meeting deadlines, and remembering appointments (Barkley et al., 2006; Kessler et al., 2005; Mannuzza et al., 1997). Both children and adults with ADHD, especially if the condition is left untreated, are more likely to use alcohol and other illegal drugs (Wilens et al., 2003). ADHD also is a risk factor for binge drinking (Niemela et al., 2006).

While the above evidence suggests the causal link by which impulsivity leads to risky drinking behaviors, it is also important to mention that there is also evidence for the opposite causal direction. The disinhibiting effects of chronic alcohol use have also been demonstrated. For example, detoxified alcoholics are impulsive when asked to complete various impulsivity questionnaires and also when tested on various behavioral tasks, and they also show greater alcohol-related response disinhibition relative to control participants (Bjork et al., 2004; Hildebrandt et al., 2004; Mitchell et al., 2005; Noel et al., 2005, 2007). Further anatomical evidence indicates that the frontal lobes, and any damage sustained in this part of the brain, contribute to impulsivity. The prefrontal lobes, in particular, play a crucial role in cognitive and behavioral control. When this region is damaged, disinhibition is observed (Jentsch and Taylor, 1999). Chronic alcohol use appears to diminish activity in the brain's frontal lobes which attenuates cognitive and behavioral control (Volkow et al., 2004). This attenuation of control may increase impulsive behaviors, including excessive alcohol consumption (Fillmore, 2003), creating a vicious circle of alcohol abuse leading to disinhibition to further alcohol abuse. Petry (2001) tested current and detoxified alcoholics and controls on the delayed discounting task described above. For delayed discounting performance, the most impulsive responding was shown by the current alcoholics, followed by the abstainers and then the controls, indicating that both the acute and chronic effects of alcohol impacts impulsivity. However, this circle may only apply to severe alcohol abuse. At this time, the role that impulsivity plays in those individuals near the start of their drinking career, such as high school binge drinkers, is not clear.

SENSATION SEEKING

Personality researchers have used slightly different terms to refer to the personality trait of impulsivity as related to substance abuse. For example, the terms behavioral approach (Gray, 1987), novelty seeking and reward dependence (Cloninger, 1987) and sensation seeking (Zuckerman, 1979, 1984) all refer to a concept similar to impulsivity (for a review see Acton, 2003). Sensation seeking may be a component of impulsivity or may be a slightly different aspect of personality. There is evidence that impulsivity and sensation seeking refer to the same concept and also evidence that these are disparate features of personality (Acton, 2003). Sensation seeking will be considered briefly since there is good evidence that as a trait it has a strong genetic component and also is a predictive risk factor for drug abuse.

Zuckerman introduced the trait of sensation seeking based on a comparative approach of humans and non-human animals. The term sensation seeking was defined as "the need for varied, novel, and complex sensations and experiences and the willingness to take physical and social risks for the sake of such experience (Zuckerman, 1979). An example might help highlight the potential distinction between the concepts of impulsivity and sensation seeking. An impulsive person could be an individual who can't help but eat all of the cookies off the plate on the table whereas a high sensation seeking person may find eating the same cookies boring. However, many impulsive individuals also like to take risks, which can make the overlap of these two constructs difficult to disentangle. To assess sensation seeking status, the Sensation Seeking Scale is a questionnaire that is often used in studies. The Sensation Seeking Scale poses various statements such as, "I would like to explore a strange city or section of town by myself, even if it means getting lost", and the participants responds true or false to each of these statements. Four different aspects of sensation seeking have been discovered including: 1) thrill and adventure seeking, 2) experience seeking, 3) disinhibition, and 4) boredom susceptibility. Zuckerman observed that in both humans and nonhuman animals, sensation seeking appeared to be genetically determined. Thus, if a parent likes to jump out of airplanes and ride roller coasters, the child of that parent is also likely to want to engage in similar experiences. Likewise, if a rat parent likes to explore novel surroundings and taste new foods, the offspring of that rat are also likely to engage in those behaviors. The tendency to seek new experiences seems to have a biological basis as several common biological correlates of sensation seeking include high levels of gonadal hormones, low levels of monoamine oxidase, and augmenting of cortical-evoked potentials (Acton, 2003).

Many studies have demonstrated the positive relationship between the frequency of alcohol and drug use and sensation seeking. The higher the individual is on the dimension of sensation seeking, the more he or she tends to use alcohol and drugs and to use more kinds of drugs (Earleywine, 1994). One explanation for this relationship is that sensation seeking represents the individual's higher degree of sensitivity to the pleasurable effects of drugs. Animal studies have also shown this relationship. For example, Suomi (1999) demonstrated that rhesus monkeys who examined an unfamiliar object dropped in their environment were also more likely to drink alcohol. Those animals that avoided or were fearful of the novel object were unlikely to drink alcohol. Similarly in rats, individual differences in exploring a novel object also were associated with amphetamine self-administration (Cain et al., 2005).

DEPRESSION AND STRESSFUL LIFE EVENTS

Depression has been extensively studied in its relation to alcohol dependence. More recent investigations have correlated depressive symptomatology with binge drinking. Bazargan-Hejazi et al. (2008) conducted a study of randomly-selected patients who were at least 18 years old and seeking emergency department care. They found that 51% of the patients reported depressive symptoms. Binge drinking, problem drinking, and drinking abuse were all related to the manifestation of depressive symptoms. Similar findings have been reported with adolescents. Simantov et al. (2000) found that depressive symptoms were associated with drinking in adolescent girls. Stressful life events were associated with increased drinking in both adolescent boys and girls. Thus, effective screening for adolescent depression and effective strategies to help students who report high levels of stress may help to reduce binge drinking rates in high school students.

BIOLOGICAL FACTORS

Some interesting biological factors have been identified from studies that have examined genetic and ethnic group differences in binge drinking college students. Luczak et al. (2001) measured binge drinking rates of Chinese, Korean, and White college students at the University of California, San Diego. They also examined variation in the aldehyde dehydrogenase (ALDH2) gene. For individuals with a deficiency in the aldehyde isozyme, a buildup of acetaldehyde occurs when drinking, since alcohol metabolism is disrupted. As such, a physical reaction occurs with drinking, called the Asian flushing response. These individuals experience several negative sensations including cutaneous flushing, heart palpitations, tachycardia, perspiration, and headache (Kitano, 1989).

The results of the Luczak et al. (2001) study identified that White students had the highest rates of binge drinking, followed by Koreans and then Chinese students. Among the Asian college students, ALDH2 status and ethnicity related to binge drinking in an additive manner. Possessing an ALDH2*2 allele (which would lead to the Asian flushing response when drinking) and being Chinese were protective factors. By contrast, being White and being Korean without an ALDH2*2 allele were risk factors for binge drinking. Therefore, ALDH2 status, as well as other cultural factors that differ in Chinese and Koreans, were associated with binge drinking in Asians.

Other genetic associations have been associated with binge drinking, separate from ALDH2 status. A genetic association of the human corticotropin releasing hormone receptor 1 (CRHR1) and binge drinking and alcohol dependence has been found. Treutlein et al. (2006) determined the allelic frequencies of polymorphisms of the CRHR1 gene in two samples: 1) in adolescents who had little previous exposure to alcohol and 2) in alcohol-dependent adults, who met DSM-IV criteria of alcohol dependence. In the adolescents, significant group differences between genotypes were observed in binge drinking, lifetime prevalence of alcohol intake, and lifetime prevalence of drunkenness. In the alcohol-dependent adults, there was an association of CRHR1 with high amounts of drinking. Thus, CRHR1 may play an important role in patterns of binge drinking and potentially alcohol dependence.

SOCIAL RISK FACTORS

Most adolescents experiment with alcohol at some point and use of alcohol escalates during the teenage years. Studies that follow adolescents over time can identify some psychosocial and behavioral risk factors for the development of binge drinking. The Growing Up Today Study has identified some of the predictors of alcohol use initiation. A prospective cohort study conducted by Fisher et al. (2007) gave self-report questionnaires to 5511 children and teenagers between the ages of 11 and 18 in 1998. The authors were particularly interested in when a teenager had their first whole drink of alcohol and when the teenager began binge drinking. They found that in their sample, 19% of teenage girls and 17% of teenage boys reported initiation of alcohol use. Some of the factors that were associated with increased likelihood to initiate alcohol consumption included older age, smoking, adults who drank in the home, peer drinking and positive attitudes towards alcohol use. Girls who ate family dinner at home every day were less likely to initiate alcohol use than were girls who ate family dinner only on some days or never. Of the subset of participants who initiated alcohol use, 29% of boys and 24% of girls engaged in binge drinking. Some of the factors that were associated with binge drinking were different for boys and girls. For boys, positive attitudes about alcohol and older age were associated with binge drinking. For the girls, positive attitudes about alcohol, underage siblings who also drank alcohol, and possession of or willingness to use alcohol promotional items were associated with binge drinking (Fisher et al., 2007). Thus, girls and boys seem to have slightly different factors associated with initiation of alcohol use and binge drinking. At least for girls, a strong family network that discourages underage drinking may play an important role in preventing or delaying the onset of binge drinking behavior. By contrast, other studies have not found gender differences in the role that the family network plays in the risk of binge drinking. In other studies, strong parental bonds and strong school bonds were protective factors against binge drinking, as were greater life satisfaction, in both boys and girls (Desousa et al., 2008; Simantov et al., 2000).

The findings from the above described Fisher et al. (2007) Growing Up Today study coincide with previous reports of psychosocial and behavioral risk factors in early adolescence as predictors of binge drinking. Griffin et al. (2000) examined several risk and protective factors during early adolescence that were associated with binge drinking in high school seniors. Using a school-based survey, students were questioned in both the seventh and twelfth grades. The authors found that heavy drinking in twelfth grade was predicted by having had experimented with alcohol or cigarettes in seventh grade. Other risk factors included having had a majority of one's friends drink and having had poor behavioral self-control during early adolescence. Similar to the findings from the Growing Up Today study, gender differences were obtained for various risk factors. In boys, positive alcohol expectancies predicted greater later heavy drinking. In girls, friends' smoking predicted later heavy drinking. Thus, prevention of early experimentation with cigarettes or alcohol may afford some protection against later binge drinking. Early experimentation with alcohol is a flag for parents and educators that the child may be at risk for further risky alcohol consumption patterns, later in high school and in college (Zakrajsek and Shope, 2006).

An impressive study completed by the Department of Defense investigated risk factors for binge drinking in young men entering the U.S. Marine Corps. A sample of 41,482 men between the ages of 18 and 20 completed questionnaires. Similar to many other studies, the authors found that early initiation of alcohol use was strongly associated with risky drinking. A 5.5 fold increased risk of risky drinking was found if the age of onset of drinking was 13 years or younger. Other risk factors included tobacco use, rural or small hometown, household alcohol abuse or mental illness, and childhood sexual or emotional abuse. Surprisingly, the authors also found that higher education, motivation to join the military for travel or adventure, and numerous close friends and relatives were also risk factors for binge drinking (Young et al., 2006). While this study only looked at young males, the findings reinforce the importance of the early onset of alcohol use as an important risk factor for later binge drinking.

Early onset drinking is also an important risk factor for the development of alcohol dependence. In recent decades, the lifetime prevalence of alcohol dependence has risen in women, but not significantly changed in men. Early age at onset of drinking has also increased in women. Grucza et al. (2008) found that early age at onset of drinking accounted for a substantial portion of change in the lifetime prevalence rates of alcohol dependence in women. Since early onset drinking is a modifiable risk behavior, it is important to limit early drinking to protect against future alcohol dependence as well as future binge drinking.

ALCOHOL ADVERTISING

Children and adolescents are exposed to a variety of forms of alcohol advertising through television, radio, online and print media. Based on findings from research about tobacco marketing and subsequent teenage smoking, investigators have examined whether exposure to and receptivity to alcohol advertising impacts rates of underage drinking. Henriksen et al. (2008) obtained data from in-class surveys on students in the sixth, seventh and eighth grades and completed a follow-up 12 months later. They found that exposure to alcohol advertising and promotions were associated with increased drinking one year later. The authors recommended that prevention programs should reduce adolescents' receptivity to alcohol marketing by limiting their exposure to alcohol ads and promotions and by increasing their skepticism about the sponsors' marketing tactics.

HIGH SCHOOL BINGE DRINKING AS A RISK FACTOR FOR COLLEGE BINGE DRINKING

High school binge drinking has been identified as a major risk factor for college binge drinking. For many individuals, drinking and binge drinking are behaviors that do not suddenly appear in college, but are familiar to the college population from their high school experiences (Wechsler et al., 1995). For example, Englund et al. (2008) followed a group of 178 adults over time from birth to age 28 years. They observed that higher amounts of alcohol consumption at age 16 coincided with later heavy drinking at ages 19 and 26 years. However, Reifman and Watson (2003) found that while high school binge drinkers were more likely to binge drink in college, it was the social network at college that was more important. Individuals involved in gregarious socializing (e.g., partying, having a social network of individuals who drank heavily) predicted college binge drinking. The authors also found that women were more likely to first adopt binge drinking in college compared to the men who may have been binge drinking in high school before the start of college.

RISK FACTORS FOR TRANSITION FROM BINGE DRINKING TO ALCOHOL DEPENDENCE

Many high school and college students drink in a binge fashion and never develop any serious problem with alcohol other than binging. As outlined in Chapter 2, binge alcohol consumption differs from alcohol abuse and dependence. However, there are subsets of young people who drink in an extreme fashion and then subsequently develop an addiction to alcohol. The results from a recent prospective study shed light on what factors predict this transition from risky drinking to alcohol dependence. Beseler et al. (2008) identified at-risk drinkers at baseline and then followed these individuals for a 10-year period. Not surprisingly, a family history of alcoholism was a strong predictor of new onsets of alcohol dependence (using the DSM-IV criteria described in Chapter 2). Furthermore, those individuals who drank to reduce negative feelings and/or who drank for social facilitation were at greater risk of alcohol dependence 10 years later if they also had a family history of alcoholism.

These findings coincide with others studies that have noted that there is a genetic predisposition at least to some, but not all, types of alcohol dependence (Capone and Wood, 2008; Kendler et al., 1992; Schuckit, 1987). Family, twin, and adoption studies have provided evidence for this genetic component of alcoholism. For example, studies of twins have shown there is greater likelihood of both identical twins having alcohol dependence than fraternal twins. Similarly, sons and daughters of alcoholics are four times more likely to develop the disorder themselves, relative to individuals whose parents are not alcohol dependent. More compelling are the findings from studies of adopted children of biological alcohol-dependent parents. Development of alcohol problems in sons and daughters is more influenced by the alcoholic biological parents than by the adoptive home environment, particularly for male children. As such, it has been estimated that genetic factors may account for up to one half of the variance in the etiology of alcoholism (Schuckit et al., 2004). Since not all cases of alcohol dependence can be accounted for by genetic factors, alcohol dependence is caused and maintained by a combination of biological, psychological, and sociological factors (Maisto et al., 2008).

CONCLUSION

A variety of risk factors have been identified that are associated with subsequent binge drinking. Some personality traits, such as impulsivity and sensation seeking, are associated with subsequent binge drinking. Mental health factors such as depression have also been associated with binge drinking in both high school students and older individuals. There appear to be genetic factors that can increase or decrease risk, including ALDH2 and CRHR1 status. Public health efforts to decrease binge drinking should target prevention of underage alcohol use, tobacco use, and limit underage exposure to alcohol advertising. Early onset drinking, in particular, is a significant risk factor for both binge drinking and alcohol dependence.

Chapter 15

ALCOHOL EXPECTANCIES AND BINGE ALCOHOL USE

One aspect of binge drinking that illustrates how different this type of alcohol consumption is from chronic drinking is the extent to which beliefs and expectations about alcohol strongly drive binge drinking behavior. Different than an alcoholic who is physiologically dependent on the alcohol and thus will experience withdrawal symptoms, such as tremors, if the person doesn't drink, the adolescent or college student who engages in binge drinking may go weeks without drinking and thus show no evidence of physiological dependence. Research on drug expectancies and beliefs has demonstrated that a person's experience with a drug, as well as his beliefs, knowledge, attitudes and other thoughts about drugs, are all part of the person and are integrated in his personality. These nonpharmacological factors, including cognitive factors such as the role of expectation, can exert extremely powerful influences on drug use and the drug experience (Vogel-Sprott, 1992; Vogel-Sprott and Fillmore, 1999). Drug expectancy is often defined as what a person expects to achieve or happen when using a drug (Maisto et al., 2008).

The role of expectation in alcohol effects has been well researched and the weight of evidence does indicate that alcohol consumption, at least in social drinkers, is heavily influenced by cognitive processes (Darkes and Goldman, 1993). Expectancies account for a larger portion of the variance in drinking behaviors among younger respondents (under age 35) compared to older social drinking respondents where expectancies play a less important role (Leigh and Stacy, 2004), perhaps since those individuals have much more personal experience with the direct effects of alcohol. MacAndrew and Edgerton (1969) published one of the early studies on expectancies. They proposed that what people learn and believe about alcohol is an important determinant of how they conduct themselves when drinking. As such, what people expect to happen when they drink can be a critical factor in determining their response to the alcohol consumed. Expectation is such a powerful force that in some cases, it may be more influential than the pharmacological action of the alcohol itself.

MEASURING THE EFFECTS OF ALCOHOL EXPECTANCIES IN THE LABORATORY

One way to investigate the role of expectation is to measure how alcohol expectancies impact behavior. In the laboratory, a researcher can give a participant various beverages (including alcohol and placebo) and observe the participant's response to each. A typical laboratory expectancy procedure was described by Marczinski and Fillmore (2005). They brought college students into the laboratory for several sessions on different days. At each session, the participant was given a beverage that they were told could contain alcohol and then they were asked to perform a task on the computer. Of importance to the study design was the fact that the participant didn't know when they would be getting alcohol. On some days, a participant would be brought a can of a soft drink. The can would be opened in front of the participant and the participant would be asked to drink the beverage. This is considered the "control" condition since neither the pharmacological effect of alcohol nor the expectancy of alcohol was involved. On another day, the researcher actually gives the participant a soft drink (nonalcoholic beverage). However, this day is different from the control condition in that the beverage is prepared out of sight of the participant. The soft drink is served in a glass that has been sprayed with alcohol (using a plant sprayer bottle). Furthermore, a few drops of alcohol have been floated on the surface of the drink just before serving. Therefore, the participant who is given this beverage thinks that the drink contains alcohol, based on the smell and the first sip. This is the placebo condition. Finally, on another day, the researcher would give the participant an actual alcoholic beverage that is served in an identical manner to the placebo condition beverage. Table 15.1 lists the conditions of importance in a typical alcohol expectancy study. Note that in some expectancy studies, the researcher may actually inform the participant that the placebo beverage contains alcohol to further convince the participant that they had consumed alcohol. This small fib is ethically acceptable since the misinformation would be disclosed to the participant when he or she is debriefed at the end of the study.

By giving a participant a "placebo" beverage and comparing the behavior of the participant to the behavior observed in the control condition, one can observe the role of expectation at work, without the actual pharmacological effect of the alcohol involved. Expectancy effects can be extremely powerful. For example, one experiment recruited undergraduate males to study their beliefs about alcohol and skills on a motor task (Fillmore and Vogel-Sprott, 1996).

The motor task involved tracking a moving target that rotated around a circle (a task known to be sensitive to the actual impairing effects of alcohol). When given a placebo drink, the individuals performed very poorly on this motor task. Their performance was so poor that they looked as if they had actually been given several standard drinks of alcohol (even though they had been given none).

Table 15.1. Typical conditions in an alcohol expectancy study

Experimental Condition	Beverage Given to the Participant	Instructions to the Participant (if given)	Factors influencing the effects of alcohol on the participant
Control	Soft drink	"This is a soft drink"	None
Placebo	Soft drink sprayed with alcoholic mist	"This is alcohol"	Expectation about Alcohol
Alcohol	Alcoholic drink	"This is alcohol"	Expectation about Alcohol + Pharmacological Effect of Alcohol

A review of various alcohol expectancy studies that were performed up to the mid-1980s found that expectancy effects were most prominent with behaviors or emotions that society associates with free expression, such as aggression, sexual arousal, and humor (Hull and Bond, 1986). Thus, if an individual thinks that drinking some alcohol will make them more aggressive, then the person will act aggressive when given a placebo, even though no amount of pharmacological effect of alcohol is involved.

MEASURING ALCOHOL EXPECTANCIES WITH QUESTIONNAIRES

Another way that alcohol expectancies have been investigated is by giving participants a questionnaire to complete that asks for their thoughts about the effects of alcohol. For example, the Alcohol Expectancy Questionnaire (AEQ) is a widely used instrument that assesses alcohol expectancies in adults and adolescents (Goldman et al., 1991). The AEQ has several different statements about the effects of alcohol. The participant is asked to read each statement and then respond whether he or she agrees or disagrees with each statement based on his or her thoughts, feelings, and beliefs about alcohol at that moment. Table 15-2 presents some sample items from typical alcohol expectancy questionnaires, including the AEQ.

Alcohol expectancies can be positive or negative. For example, if a participant agrees with the item that "A drink or two makes the humorous side of me come out," the individual has a positive expectation that alcohol increases humor. However, some alcohol expectancies are negative. For example, positive endorsement of an item such as, "I'm more likely to say embarrassing things after drinking", indicates that the participant has a negative expectation that alcohol increases the likelihood that he or she might make inappropriate verbal statements.

Where do these alcohol expectancies come from? A person's expectancies can be based on previous experiences with alcohol and its effects. Thus, if a teenager drinks alcohol and his friends say that he was so funny when he was drinking, this experience will lead to the development of an expectation that alcohol increases humorous actions. However, direct experiences may not be necessary to develop expectations about alcohol. Indirect experiences (i.e., social learning or watching others) have also been shown to impact the development of alcohol expectancies.

Table 15.2. Example items used in questionnaires to assess alcohol expectancies

Participant Instructions: Respond to these items according to what you personally believe to be true about alcohol		
	Agree	Disagree
Alcohol makes me feel happy.		
Drinking adds a certain warmth to social occasions.		
A drink or two makes the humorous side of me come out.		
I find that conversing with members of the opposite sex is easier for me after I have had a few drinks.		
A few drinks make me feel less shy.		

For example, watching friends or relatives use alcohol, watching television shows where alcohol is consumed, and viewing advertising that involves alcoholic products are all ways in which alcohol expectancies can develop.

ALCOHOL EXPECTANCIES AND DRINKING HABITS

Correlational studies have highlighted the finding that alcohol expectancies gathered from questionnaires are associated with the initiation and maintenance of alcohol consumption (Darkes and Goldman, 1993). Regardless of whether an individual is a light social drinker or an alcoholic, positive alcohol expectancies have been shown to relate to drinking levels in adolescents and adults (Brown, 1985; Brown et al., 1985; Christiansen et al., 1982; Christiansen and Goldman, 1983; Connors et al., 1986; Cooper et al., 1988; Mann et al., 1987; Mooney et al., 1987). For example, Rather et al. (1992) reported that heavy drinkers have different alcohol expectancies compared to their lighter drinking peers. Heavier drinkers associate arousal and energy with drinking, whereas light drinkers associate sedation with drinking.

CHILDREN AND ALCOHOL EXPECTANCIES

It is fascinating that even very young children have expectations about alcohol use. Once it became apparent that alcohol expectancies may be driving alcohol consumption patterns in adults, it was imperative to determine whether alcohol expectancies lead to the initiation of alcohol use in children and adolescents. Children and adolescents (before any direct experience with alcohol consumption) have detailed expectations about how alcohol will affect behavior (Christiansen et al., 1989; Christiansen and Goldman, 1983; Miller et al., 1990). Furthermore, alcohol expectations have been found to predict drinking onset as well as the onset of problem drinking once adolescents have begun to drink (Christiansen et al., 1989).

The development of alcohol expectancies in children appears to follow a predictable pattern where young children associate alcohol consumption with negative outcomes and older children associate alcohol with positive consequences. Late childhood may be a critical transition period where expectancies about alcohol use shift from negative to positive. For example, a young child may hold an expectation that alcohol makes an adult act silly (negative expectancy) whereas an older child shifts to an expectation that alcohol is associated with parties and fun (positive expectancy). Dunn and Goldman (1998) assessed expectancies and self-reported drinking in a large sample of 2,324 school children from grades 3, 6, 9, and 12 from a large suburban school district. They found that the older children, and especially those children who had started drinking, endorsed more positive and arousing alcohol expectancies as they got older (see also Cameron et al., 2003). Children who had not begun drinking associated undesirable effects with alcohol (Dunn and Goldman, 1998). Since positive alcohol expectancies are associated with the onset and maintenance of drinking, and early onset drinking in children is associated with binge drinking and alcohol dependence, any effort to delay the shift from negative to positive alcohol expectancies should improve health and limit binge drinking.

ALCOHOL ADVERTISING AND EXPECTANCIES

Part of the transition the children make from holding largely negative alcohol expectancies to holding positive alcohol expectancies may be due to watching peers, other adolescents, parents and other adults use alcohol. However, alcohol advertising may also be part of this transition process. Children are exposed to advertising from a variety of sources including print, television and online sources. Some typical sources include in-store beer displays, exposure to magazines with alcohol advertisements, beer concession stands at sports or music events and television ads promoting various alcoholic beverages. All alcohol advertising is intended to increase positive expectancies toward a brand of beverage (e.g., Brand X of beer is associated with parties, lots of beautiful people and fun). A variety of researchers have examined the relationship between exposure to various forms of alcohol advertising, alcohol expectancies and actual drinking in adolescents. In general, increased exposure to alcohol advertising in children is associated with increased favorable attitudes and beliefs about drinking (positive alcohol expectancies), increased intentions to drink as an adult, and increased underage drinking (Ellickson et al., 2005; Fleming et al., 2004; Grube and Wallack, 1994). Efforts to limit exposure to alcohol advertisements to only those individuals who are of age to drink would be helpful in combating these positive expectancies about alcohol.

ALCOHOL EXPECTANCIES AND 21ST BIRTHDAY DRINKING

College students appear to have an expectation about the role alcohol plays in the celebration of an individual's 21st birthday. This expectation sometimes leads to extreme cases of binge drinking. While many college students have long been drinking alcohol before they were legally able to do so, there is a trend to have a grand celebratory drinking event to mark the actual legal transition to being able to drink at the age of 21. Unfortunately, the expectation of how to celebrate a 21st birthday has become a dangerous one. Many individuals now even claim to mark their 21st birthday with consumption of 21 drinks. This level of drinking is extremely dangerous. Using a typical 180 pound male as an example, the consumption of 21 beers over an eight hour period of time would result in a peak blood alcohol concentration of more than .30 g% (almost four times the legal limit of driving of .08 g%). This blood alcohol concentration would most likely result in loss of consciousness due to alcohol poisoning which would put the individual at risk of choking on his own vomit, and death by asphyxiation.

One recent study examined this phenomenon of having 21 drinks to celebrate a 21st birthday using a survey given to over 2000 college students who had celebrated a 21st birthday (Rutledge et al., 2008). The authors found that 83% of those surveyed stated that they drank alcohol to celebrate their 21st birthday. Some respondents who claimed that they had consumed 21 drinks on that day were individuals who also stated that they had previously been drunk on more than one occasion or belonged to a fraternity or sorority. Half of the survey respondents also stated that they had drunk more on that birthday than they had ever drunk before in their life. While the generalizability of this study is uncertain, since it was only completed at one university and the reports of drinking behavior were based on self-reported survey responses, the behavior in question is disconcerting. Having a goal of drinking 21 drinks on a 21st birthday is a potentially deadly expectation about how alcohol should be consumed on that day. It also appears that this expectation is based somewhat on inaccurate perceptions about the amount of drinking actually occurring on that birthday. A study by Neighbors et al. (2006) found that students overestimated the amount of peer drinking that occurs at 21st birthday celebrations. These misperceptions about peer drinking were not limited to birthday celebrations. Tailgating at college football games is also a social context where a great deal of binge drinking occurs. The same authors also found that students overestimated the amount of alcohol that the typical tailgater consumed. However, they also underestimated the percentage of tailgaters who drank some alcohol, suggesting that tailgating is a social context at college where drinking is the norm, not the exception.

CHALLENGING ALCOHOL EXPECTANCIES

A certain amount of evidence is consistent with a causal interpretation of the role of alcohol expectancies in the initiation and maintenance of alcohol consumption patterns (Darkes and Goldman, 1993). Alcohol expectancies can lead to binge drinking, especially if the expectancies that are held are inappropriate and potentially dangerous (e.g., expectancies about how much alcohol should be consumed on a 21st birthday celebration). Therefore, researchers have been interested in determining whether there are ways to challenge currently held alcohol expectancies to moderate drinking behaviors. Moderate to heavy drinking college students may benefit from a challenge to their currently-held beliefs about alcohol.

To determine whether expectancies can be modified in a way that also changes behavior, Darkes and Goldman (1993) selected male college students who drank in a moderate to heavy fashion and randomly assigned them to one of three groups in a pre-post design. One group was exposed to an Expectancy Challenge condition (this condition was designed to manipulate expectancy levels). Another group was given traditional information about alcohol and the final group was the control condition, given no information about alcohol. For the Expectancy Challenge condition, participants had their alcohol beliefs challenged by having them consume beverages (both alcohol and placebo) so that they could identify in themselves and observe in others the effects of alcohol. The challenge procedure was designed to undermine and disrupt subject's existing associations between drinking and expected behavioral outcomes. In particular, the researchers wanted to show participants that the increased sociability and increased sexual behavior often thought as benefits of alcohol consumption were actually largely due to expectations and not the pharmacological properties of alcohol. For example, if a person acted and felt more sociable after a drink and then was informed that the drink was actually a placebo, this experience should challenge the participant's previously-held belief that alcohol increases sociability.

The expectancy challenge procedure not only changed expectancies, but it also changed behavior. Participants in the expectancy challenge condition exhibited decreases in measured positive alcohol expectancies in the post test compared to the pre test. Similarly, participants in the expectancy challenge condition also significantly decreased their actual drinking. Significant decreases in alcohol expectancies and actual drinking were not observed in the alcohol knowledge condition. Therefore, just teaching individuals that alcohol does not increase sociability does not change beliefs and/or behavior. Instead, expectancies need to be challenged in a more dramatic fashion in order to change expectancies and also change behavior.

While the traditional expectancy challenge procedure may be effective in changing drinking patterns of drinkers 21 years and older, many binge drinkers are underage. Could expectancies be challenged in individuals who are below drinking age by using observational techniques, since the presentation of alcohol would not be legal? The importance of challenging expectancies in children and adolescents has grown as it appears that alcohol expectancies undergo a critical phase of development during late childhood and early adolescence (Christiansen et al., 1989; Miller et al., 1990). Kraus et al. (1990) demonstrated that the development of alcohol expectancies can be slowed in children. Using elementary school children, the authors found that the development of alcohol expectancies were slowed by films that attempt to undermine expectancies. By using adaptations to the traditional expectancy challenge procedure, such as presentation through media, and using age-appropriate material, a change in beliefs can be acquired.

Austin and Johnson (1997) also found that children may respond more readily to media materials in expectancy challenge procedures, compared to adults who may be more resistant to change their beliefs about alcohol and its effects. They had third-grade children view a videotape about television advertising, as well as video clips of alcohol ads. The viewings were followed by a discussion pertaining to alcohol advertising. This process was sufficient to change alcohol expectancies among the third-graders. Children exposed to this process had decreased expectations of the positive consequences from drinking alcohol and decreased likelihood to choose an alcohol-related product. In a similar study, Cruz and Dunn (2003) successfully altered alcohol expectancies in fourth-grade children. Following their expectancy intervention, the children were more likely to associate alcohol use with negative and sedating consequences and were less likely to associate alcohol with positive and arousing outcomes. Thus, expectancy challenge procedures appear to be effective means to change beliefs and thoughts about alcohol use. This is especially important in young children, since late childhood may be a critical transition period in expectancies about alcohol use.

CONCLUSION

Alcohol expectancies are defined as what a person expects to occur when using alcohol. Cognitive factors, such as the role of expectation, can exert extremely powerful influences on alcohol use. Alcohol expectancies can be positive (e.g., alcohol makes people more sociable) or negative (e.g., alcohol makes people act inappropriately). There appears to be a developmental change whereby young children often hold negative expectancies about alcohol use whereas older children hold more positive expectancies about alcohol use. Since positive alcohol expectancies are associated with the onset and maintenance of drinking, and early onset drinking in children is associated with binge drinking and alcohol dependence, any effort to delay the shift from negative to positive alcohol expectancies should improve health and reduce binge drinking. Exposure to alcohol advertising increases positive alcohol expectancies. Underage individuals should therefore have limited exposure to this type of advertising. Expectancy challenge interventions appear to be successful in changing alcohol expectancies and may help moderate binge drinking in adolescents and college students.

Chapter 16

TRENDY COCKTAILS:
MIXING ENERGY DRINKS AND ALCOHOL

Alcoholic drink preferences in college students have shifted in the past decade, with trends in consumption leaning toward alcohol mixed with energy drinks (AmED). Specific examples include Red Bull and vodka or mixed cocktails with names such as Jager Bomb (1/2 can Red Bull mixed with 1 oz. Jagermeister). Energy drinks (e.g., Red Bull, Full Throttle, Monster, Rock Star, Amp, No fear, Adrenaline Rush, etc.) are beverages marketed as providing increased energy using a combination of caffeine and other plant-based stimulants (e.g., guarana) and amino acids (e.g., taurine). High-caffeine soft drinks have existed in the United States since the 1980s, beginning with Jolt Cola. However, energy drinks began being marketed as a separate beverage category from soft drinks, with the introduction of Red Bull, an Austrian product in 1997. Energy drink consumption and sales have since exploded with a 516% inflation-adjusted increase from 2001 to 2006 alone. In 2006, there were as many as 500 new energy drink products introduced worldwide, but Red Bull still remains the largest manufacturer with 43% of the market (Simon and Mosher, 2007).

ENERGY DRINKS

The consumption of energy drinks is a popular practice among children, adolescents, and young adults in a variety of situations. Mintel, a leading market research company, has provided data quantifying the explosion in the energy drink market. They valued the energy drink retail market at $4.8 billion, a growth of over 400% from 2003. Similarly, the number of energy drinkers has also escalated. In 2003, only 9% of adult respondents to Mintel's survey said they drank energy drinks. In 2008, 15% did. However, the growth in the adult market pales to that observed with teens. In 2008, 35% of teenagers regularly consumed energy drinks, in comparison to 19% of teens in 2003 (Mintel, 2008). Energy drink use may be most common in college students. One recent survey of undergraduate students at a public university in New York indicated that 39% of undergraduates had consumed at least one energy drink in the past month. A gender difference in use of energy drinks was also observed with males stating more frequent use (2.5 drinks/month) compared with females (1.2 drinks/month) (Miller, 2008).

The rise in use of energy drinks might reflect user observations that these drinks increase energy. In a few studies, ingestion of energy drinks alone has been shown to improve a variety of physical and mental aspects of performance in both humans and animals. For example, psychomotor performance (e.g., motor reaction time), driving performance, concentration, memory, verbal performance, subjective sensation of alertness, physical vigor, physical performance and mood have all been improved following the consumption of energy drinks (Alford et al., 2001; Ferreira et al., 2004; Horne and Reyner, 2001; Seidl et al., 2000; Warburton et al., 2001). Even though energy drinks contain a variety of compounds, the high caffeine content of these drinks appears to drive the stimulant effects of these beverages (Ferreira et al., 2006; McCusker et al., 2006). For example, Coca-Cola Classic contains 2.9 mg of caffeine/fl oz., whereas Red Bull contains 9.2 mg of caffeine/fl oz. In addition, soft drinks cans contain 12 ounces of beverage whereas energy drinks are often served in 16 ounce containers for many popular brands. Interestingly, the U.S. Food and Drug Administration (FDA) limits the amount of caffeine to 65 milligrams per serving of a food or beverage. However, the FDA currently does not regulate energy drinks, which is why they can contain as much as 300 milligrams of caffeine in a single serving (Fox News, 2007; McCusker et al., 2006). Table 16-1 lists several popular brands of soft drinks and energy drinks with their caffeine content.

Table 16.1. Examples of the caffeine content of soft drinks and energy drinks sold in the United States. Sources of caffeine information include caffeine information listed on beverage cans, company website product information retrieved September 2008 from the Coca-Cola Co., PepsiCo Inc., Dr. Pepper/Seven Up Inc., Beverage Partners Worldwide, Monster Beverage Co., Redux Beverages, Rockstar, and the article *How much caffeine is in your daily habit?* **Retrieved from http://www.mayoclinic.com/health/caffeine/AN01211**

Brand	Serving Size in Can (ounces)	Total Caffeine (mg)	Caffeine (mg)/oz. of beverage
Soft Drinks			
Coca-Cola Classic	12	35	2.9
Pepsi	12	38	3.2
Diet Coke	12	47	3.9
Diet Pepsi	12	35	2.9
Mountain Dew	12	54	4.5
Diet Mountain Dew	12	54	4.5
Code Red Mountain Dew	12	54	4.5
7Up	12	0	0
Sprite	12	0	0
Sprite Zero	12	0	0
Sunkist Orange Soda	12	41	3.4
AandW Creme Soda	12	29	2.4
Barq's Root Beer	12	23	1.9
Diet Barq's Root Beer	12	0	0
Dr. Pepper	12	41	3.4
Diet Dr. Pepper	12	41	3.4
Fanta	12	0	0
Mello Yello	12	53	4.4

Brand	Serving Size in Can (ounces)	Total Caffeine (mg)	Caffeine (mg)/oz. of beverage
Tab	12	47	3.9
Energy Drinks			
Red Bull	8.3	76	9.2
AMP	16	143	8.9
Enviga	12	100	8.3
Full Throttle	16	144	9.0
Full Throttle Fury	16	144	9.0
Monster Energy	16	160	10.0
No name (formerly Cocaine)	8.4	280	33.3
SoBe Adrenaline Rush	16	152	9.5
SoBe No Fear	16	174	10.9
Rockstar	16	160	10.0
Vault	8	47	5.9
NOS	16	250	15.6
Bawls	8.45	56	6.7
Crunk	8	100	12.5
Redline	8	250	31.3
V	8.45	78	8.6
Spike Shooter	8.45	300	35.7
Sparks (also 6% alcohol per vol.)	16	87	5.4
Soft Drink/Energy Drink Mixes			
Mountain Dew MDX	12	71	5.9
Coca-Cola Blak	12	70	5.8
Pepsi Max	12	69	5.6
Tab Energy	8	72	9.0
Jolt Cola	12	72	6.0
Jolt Endurance Shots	2	150	75.0

Table 16-1 illustrates the high caffeine content found in many energy drinks on the market. Not surprisingly, there have been many instances of high school students reporting side effects from these beverages. Caffeinism (caffeine intoxication) generally increases in likelihood when daily consumption of caffeine exceeds 600 milligrams, with symptoms including shakiness and headache. Consumption of 1000 mg or more can result in more severe symptoms including muscle twitching, rambling thought and speech, cardiac arrhythmia, periods of inexhaustibility, and psychomotor agitation (American Psychiatric Association, 1987). The likelihood of death is low since the lethal dose of caffeine is approximately 10 grams for adults and 100 mg/kg for children (Leonard et al., 1987), the equivalent of 200 cans of a soft drink. There are anecdotal reports of teens experiencing caffeine intoxication from the heavy use of energy drinks. In a 2008 news report, four middle school students from Florida were sent to the emergency room since they were experiencing heart palpitations and sweating following the consumption of the energy drink, Redline. Several high school students from Colorado became ill after drinking the energy drink, Spike

Shooter (Parker-Pope, 2008). Table 16-1 reveals that both Redline and Spike Shooter have higher caffeine content than many other energy drinks on the market.

ALCOHOL MIXED WITH ENERGY DRINKS (AmED)

Young partygoers have become enamored with using energy drink beverages as a mixer for their alcoholic drinks, presumably for the purpose of reducing the depressant effects of alcohol and thus allowing them to drink and party longer (Ferreira et al., 2006; Marczinski and Fillmore, 2006; Oteri et al., 2007). The increased popularity of heavily caffeinated alcoholic beverages was quickly identified by the beverage industry in North America. In 2005, several "energy" beers and malt beverages were introduced such as Miller-Coors' Sparks, Labatt's Shok, Molson's Kick, and New Century Brewing's Moon Shot. Physicians have warned of the potential health implications of these new heavily caffeinated cocktails, such as an increased risk of dehydration (American Medical Association, 2003). Similarly, the alcoholic beverage manufacturers of these premixed AmED beverages have come under fire. Anheuser-Busch agreed in June of 2008 to stop selling caffeinated alcoholic beverages (including their popular products Tilt and Bud Extra) after an investigation by New York and other states stated that the company was illegally marketing the drinks to young people. In a statement, New York Attorney General Andrew Cuomo said "Drinking is not a sport, a race or an endurance test. Adding alcohol to energy drinks sends exactly the wrong message about responsible drinking, most especially to young people." Investigations into other producers of alcoholic energy drinks continue (Spector, 2008).

As for college students, 1/3 of regular energy drink consumers are in the 18 to 24-year old demographic group, a group also known to have the highest rates of binge drinking and impaired driving (Flowers et al., 2008; Wechsler et al., 1995, 1998, 2000). Despite the huge rise in the popularity of these beverages, there are few published studies on the rates of consumption of energy drinks and AmED in college students. Malinauska et al. (2007) surveyed college students attending a state university in the Central Atlantic region of the U.S. and reported that 51% of students had consumed at least one energy drink in the past month. Of these users of energy drinks, 54% stated that they consumed them with alcohol. These high rates of AmED use have been reported in other countries. In another recent study, 48% of medical students from the University of Messina in Italy reported using AmED at least once in the past month (Oteri et al., 2007).

However, reported rates of energy drink and AmED use are variable. Several reasons may account for this fact. First, there has been a fairly dramatic rise in actual use, at least according to reports of energy drink sales and anecdotal reports from bars and restaurants that energy drink cocktails are becoming an increasing proportion of their alcohol sales. Second, the way investigators have chosen to collect data on use has varied considerably from study to study. Some studies report rates of AmED use only for participants who are energy drink users, whereas other studies report AmED use only for alcohol drinkers. These different data presentation approaches makes comparisons between studies difficult. For example, a fall 2006 survey of college students from North Carolina indicated that approximately ¼ of college student alcohol drinkers were also consuming AmED (O'Brien et al., 2008). By contrast, Malinauska et al. (2007) surveyed college student users of energy drinks, and found that 54% of energy drink users also stated that they consumed them with alcohol. In addition, rates of AmED use are unknown in underage drinkers, since the previous studies only questioned legal drinkers.

MOTIVATIONS FOR USING AMED

There is very little knowledge about what motivates high school and college students to use AmED. On the surface, it would appear that the price of these beverages would be a strong deterrent for use, considering that most energy drinks themselves exceed $3 a can in convenience stores and many bars charge between $8 and $10 for an energy drink cocktail. Although to date there has been no published study of North American drinkers' motivations for using AmED, a small survey of college students in Brazil asked participants to report why they used these new drinks. The authors reported that 76% of their sample used AmED. The participants stated that they mixed energy drinks with vodka, whiskey or beer. Of those users of AmED, 38% reported that the consumption of AmED increased happiness, 30% reported euphoria from these drinks, 27% reported uninhibited behavior from these drinks, and 24% reported increased physical vigor.

It is interesting that the users reported AmED use to result in uninhibited behavior, since uninhibited behavior has been associated with these new drinks. Reports from Ireland and Germany have implicated these drinks in assaults and automobile accidents, respectively, suggesting that the combination of a stimulant and alcohol impairs the ability to correctly assess level of intoxication and the ability to drive more than alcohol intoxication alone (Riesselmann et al., 1996; Tormey and Bruzzi, 2001). One U.S. private-practice attorney has been quoted as saying that 70% of his cases for DWI defendants ages 20 to 27 involved mixing alcohol with energy drinks (13WHAM-TV Rochester NY, 2005).

Energy drinks and AmED have been available to consumers in European countries for a longer period of time than in the United States. Associated problems (e.g., increased accidents rates) have accumulated and Denmark, France, and Norway have placed bans on the sale of Red Bull, citing health concerns. Despite other countries' concerns with energy drinks, they remain extremely popular with young people all over the world. One reason that AmED may be appealing to high school and college drinkers is that energy drinks have an extremely sweet taste, due to their high sugar content (e.g., one 250 ml can of Red Bull contains 27 grams of sugar). In studies that have asked underage drinkers what form of alcohol they had first consumed when alcohol consumption was initiated, alcohol use in humans is usually initiated using sweetened alcohol solutions (Fromme and Samson, 1983; Margulies et al., 1977). Furthermore, novice adolescent drinkers also use highly concentrated alcohol solutions (e.g., liquor) as the most widely-used form of alcoholic beverage (Eaton et al., 2006). Therefore, it could be that the adolescents favor the sweet taste of the energy drink to mask the high alcoholic content. Energy drinks are a very sweet tasting mixer that can be mixed to a palatable-tasting result with high alcohol content liquors, such as vodka.

LABORATORY INVESTIGATIONS OF AMED

There have been surprisingly few laboratory investigations into the subjective and objective reactions to the consumption of energy drinks alone and in combination with alcohol. However, the few that have been published have found that AmED consumption often reduces the depressant effects and/or increases the excitatory subjective effects of alcohol.

In other words, participants will report feeling less sleepy and/or more alert with AmED compared to alcohol alone. For example, one study with Brazilian college students evaluated the effects of AmED (vodka + Red Bull) compared with alcohol or energy drink alone. The authors reported that ingestion of AmED was associated with reduced perception of headache, weakness, dry mouth, and feelings of motor impairment compared to alcohol alone.

Participants were similarly impaired by AmED and alcohol alone on two objective measures: motor coordination and visual reaction time. The ingestion of the energy drink did not alter the breath alcohol concentrations (Ferreira et al., 2006). This study replicated another study that found that AmED and alcohol consumption resulted in effects similar to a maximal effort test (Ferreira et al., 2004).

These results are consistent with the more established literature on the findings of mixing caffeine with alcohol. Coadministration of caffeine with alcohol often reduces participants' perceptions of alcohol intoxication compared with the administration of alcohol alone. However, coadministration of caffeine with alcohol does not counteract all of the impairing effects on behavior, particularly the impairing effects of alcohol on impulse control (Marczinski and Fillmore, 2003a, 2006).

With AmED consumption, a potentially worrisome situation may arise when subjective intoxication is reduced yet behavioral impairment still exists. Subjective perceptions of intoxication level may function as feedback for an individual to terminate their drinking episode or to avoid potentially hazardous activities (e.g., driving). The mix of poor impulse control and reduced feelings of intoxication is the combination that may make AmED consumption riskier than alcohol consumption alone. Adding another layer of risk is that binge drinkers have poorer impulse control in general, which gets exacerbated when they are drinking (Marczinski et al., 2007).

AmED Consumption and Risky Behavior

AmED consumption may lead to riskier behavior than alcohol consumption alone. O'Brien et al. (2008) found that AmED use was associated with heavy episodic drinking (i.e., binge drinking) and more episodes of getting drunk compared to alcohol use alone. The individuals who consumed AmED reported getting drunk on average 1.4 times per week whereas the alcohol drinkers (non-AmED) reported getting drunk on average 0.7 times per week.

Students who consumed AmED were at greater risk for alcohol-related consequences, such as driving while impaired, riding with an intoxicated driver, and requiring medical treatment for an injury following drinking, even after adjusting for the amount of alcohol consumed. AmED users were also more likely to report being taken advantage of sexually or taking advantage of another person sexually compared to non-AmED users (O'Brien et al., 2008). However, studies such as this rely on cross-sectional data. This limits the ability to truly make causal statements about whether AmED leads to risky behavior. However, it does appear that AmED consumption is, at the least, strongly associated with the incidence of risky behaviors.

AMED MAY LEAD TO ESCALATION OF DRINKING

No published research on humans thus far has provided empirical evidence confirming the anecdotal reports that AmED use reduces the sedative effects of alcohol, allowing an individual to drink more and longer. However, there is some animal evidence that suggest this may be the case. A study with rats demonstrated that caffeine can increase the stimulant properties of low doses of alcohol, and increase the voluntary consumption of alcohol (Kunin et al., 2000). Similarly, coadministration of another stimulant drug, nicotine, also increased alcohol consumption in social drinkers (Harrison and McKee, 2008). Therefore, it is a possible that mixing a stimulant, like an energy drink, may be leading to more binge drinking.

Another line of evidence that AmED may lead to an escalation of drinking is that laboratory studies have found that high caffeine doses speed up the development of alcohol tolerance. For example, Fillmore (2003) had social drinkers participate in a multiple session study whereby participants were randomly assigned to one of three conditions. Participants either received alcohol (0.65 g/kg), caffeine (4 mg/kg) or both over several sessions. A 4 mg/kg dose of caffeine is 280 mg of caffeine for a typical 70 kg individual which is the equivalent of consuming 4 cans of the energy drink Red Bull or 2 cans of the energy drink Amp. On each session of this study, participants performed a psychomotor task under the assigned dose. On the final session, the author tested tolerance by giving a challenge dose of alcohol (0.65 g/kg) and observing performance on the task. Results showed that a history of combined alcohol and caffeine administration increased alcohol tolerance compared with an exposure history of either alcohol or caffeine alone. In other words, participants who had been repeatedly receiving the alcohol mixed with caffeine were less impaired by the alcohol alone on the final test. When tolerance develops, higher doses of alcohol might be needed to reinstate the initial effect, which is why alcohol tolerance has become recognized as a factor that may contribute to alcohol abuse and dependence by encouraging the use of escalating doses (American Psychiatric Association, 1994). As such, the Fillmore (2003) study provides some preliminary evidence that AmED may be escalating tolerance which may lead to harmful drinking patterns. In summary, energy drink and AmED consumption appears to be common among college students and combining energy drinks with alcohol may increase the risks associated with alcohol consumption (Ferreira et al., 2006; Marczinski and Fillmore, 2006; O'Brien et al., 2008), including the risk of binge drinking. However, considerable research is still needed on this issue.

CONCLUSION

There has been a dramatic rise in the consumption of energy drinks in all segments of the population (children to adults). Alcohol mixed with energy drinks (AmED) has become the beverage of choice for many underage and college drinkers. Users frequently report greater pleasure and less sleepiness when using AmED, compared to alcohol alone. There have been few studies of rates of AmED use and the acute effects of AmED on objective and subjective measures in the laboratory. However, although the existing literature demonstrates some similarities between the effects of alcohol alone and when mixed with energy drinks, the studies also demonstrate significant differences. However, the existing research has reported a dissociation in reaction to AmED compared to alcohol alone. For example, even though the subjective perceptions of some symptoms of alcohol intoxication are less intense after AmED consumption, compared with alcohol alone, these reduced effects have not been detected in objective measures of behavior, such as motor coordination and reaction time. It is possible that reduced feelings of sedation following drinking may lead to further drinking and increase the likelihood that a binge will occur. The paucity of studies on AmED is problematic for adequately informing the public about the risks of these drinks. Since AmED consumption is becoming relatively commonplace, the provision of warning labels on energy drinks regarding the danger of consuming these beverages with alcohol may be important. Likewise, high school and college-based programs to reduce high-risk drinking should consider including information about the risks of mixing alcohol with energy drinks as part of an overall program to reduce binge drinking and its consequences.

Chapter 17

PREVENTION AND INTERVENTION STRATEGIES

With every death of a high school or college student due to binge drinking, there is a public outcry of what should have been done to prevent such a senseless tragedy. Even the most self-reliant children seem not to be immune to such a fate. The death of Danny Reardon is good example. Danny's father had been fully confident that his son, age 19, would be capable of dealing with the pressures of college life and taking care of himself. His son had spent the past nine months after his high school graduation traveling across Europe. As a freshman, Danny had only been in college for a few months when his father, a dentist in Washington, D.C., received a call from University of Maryland police that his son had taken part in a fraternity drinking ritual and was unconscious. According to police reports and court records, fraternity members had put Danny on the sofa and took his pulse and had taken turns watching him. Danny had stopped breathing early in the morning. Students at the fraternity house had called for an ambulance around 3:30 a.m., but by the time he reached the hospital, his brain had ceased functioning. He died six days later. In this case, the tragedy of Danny's death was not just limited to the child who died. Danny's father sued the fraternity with claims in the civil suit stating that the students in the fraternity failed to provide any reasonable assistance and appropriate medical care to Danny until it was far too late. The Reardon family reached an undisclosed settlement with some of the students and Phi Sigma Kappa, which denied liability (Davis and DeBarros, 2006).

With the rise of binge drinking over the last 20 years, there has been a subsequent realization among professionals who regularly come in contact with this problem that an ounce of prevention is worth a pound of cure. Drinking and related problems on college campuses have reached near epidemic levels (Walters et al., 2000). As such there have been multiple approaches taken to get young people to abstain or at least to drink in a more moderate fashion. This chapter covers and evaluates the success of some of these approaches.

APPROACHES BY HIGH SCHOOLS

The most widely-used curriculum in schools to reduce drug and alcohol use is the Drug Abuse Resistance Education (DARE) program. In the United States, approximately 26 million children participate in a DARE program and another 10 million children in another 43 countries around the world do so also. In the U.S., 75% of the nation's school districts have adopted DARE as their primary alcohol and drug prevention program, with an estimated cost of more than a billion dollars in 2001. It is important to note that the DARE program does not try to specifically reduce binge drinking in adolescents. Instead, it is a more generalized program that is designed to give children the skills needed to avoid involvement in drugs, gangs and violence. Alcohol use plays a part, albeit small, in a curriculum that tries to address many disparate types of problems, including prescription drug abuse and cyber bullying. DARE is typical of the resistance-skills training approach to drug use prevention. The focus of any resistance-skills training program is to help the child understand alternatives to drug use, and to develop the skills needed to recognize and resist peer pressure.

The DARE program was founded in 1983 in Los Angeles, CA, when the Los Angeles Police Department was dealing with an escalation in drug abuse problems. Since the police experienced the ravages of illicit drug use, who better than a police officer to lead a school-based program to prevent initiation of drug use. Today, police officers with special training in child development lead a series of classroom lessons designed to help students build skills according to the DARE approach, which is: 1) D – Define problems and challenges, 2) A – Assess available choices, 3) R – Respond by making a choice, 4) E – Evaluate their decisions. The DARE program can be used with school children from kindergarten through 12th grade. For the most part, however, the DARE program historically targets fifth and sixth grade students before they enter junior high school. The curriculum includes 10 modules that last approximately 60 minutes. The modules focus on topics such as refusal skills, critical thinking, and alternatives to substance use. Students also sign a pledge not to use drugs or join a gang. The goal is that the students finish DARE ready to resist peer pressure and live productive drug and violence-free lives (Maisto et al., 2008; see also www.dare.com).

Despite the pervasive use of DARE in most schools, efficacy studies of the DARE program have been disappointing. Most studies have found that not only does the DARE program not reduce drug or alcohol use in its participants, rates of illicit drug use have actually been higher in DARE participants compared to those who had not gone through the program (Botvin and Griffin, 2003; Ennett et al., 1994; Lynam et al., 1999; West and O'Neal, 2004). For example, Lynam et al. (1999) followed students who participated in DARE while in the sixth grade and determined that, 10 years later, the students did not have more successful outcomes than students not exposed to DARE. In 2001, the U.S. Surgeon General placed the DARE program in the category of "Does Not Work". The only positive aspect to the program that has been clearly demonstrated is that children who participate in DARE have more favorable attitudes toward the police. Despite the efficacy studies that question the utility of the DARE program, DARE is still very popular and well-funded in the U.S. It also not surprising that the DARE program has not impacted binge drinking rates in adolescents, as underage alcohol use is only a very small component of the DARE curriculum.

Another example of an injury prevention strategy targeting alcohol use in high school students is the P.A.R.T.Y. Program (Prevent Alcohol and Risk Related Trauma in Youth). This Canadian Program was first developed in 1986 at Sunnybrook and Women's Hospital in Toronto, and has subsequently grown to include programs in 68 Canadian communities, as well as 4 in the US (3 in Colorado and 1 in Wisconsin) and 2 in Australia. The P.A.R.T.Y. Program is a single day, in-hospital injury awareness and prevention program targeted at high school aged students. It has been described as a 'fast paced tell it like it is' injury prevention program that educates teenagers about the outcomes of making poor choices (see www.partyprogram.org). In some of the larger centers, programs are run 2-3 times per week for groups of 35-40 students.

Students interact with a multi-disciplinary team of health professionals that includes paramedics, police, nurses, physicians and social workers. These professionals help students follow the course of their decisions from injury occurrence through transport, treatment, rehabilitation and re-integration using previous and current patients and touring the hospital facilities, including the trauma bay and the intensive care unit. The students are informed about: basic anatomy and physiology, the mechanics of injury, the effect of drugs or alcohol on decision making, risk assessment, coordination and concentration, and the effect of trauma and injuries on families, finances and future plans. The goal is for the students to recognize potential injury situations, to make injury prevention oriented choices and adopt behaviors that minimize risk. Most participants report the experience as an eye-opening one and is a must attend for all young people.

The P.A.R.T.Y. program has been shown to effectively improve youth driving-related attitudes. Using a validated measure of "risk-taking attitudes among young drivers" questionnaire, 11 dimensions of risk-taking attitudes were measured both before and after attending a P.A.R.T.Y. Program. Significant improvements in risk-taking attitudes were observed across all demographics after attending the P.A.R.T.Y. program (Arruda et al., 2007). Further evaluation of this program is needed.

The seriousness of the impaired driving problem in high schools has led to some extreme approaches that have been championed by some but criticized by others as unethical and possibly psychologically harmful. One example is the "Every 15 Minutes" program has been described as an intensive two-day program designed to teach students about the hazards of driving under the influence of alcohol. This mock exercise involves removing a student from class every 15 minutes and then having a police officer inform the classroom that the removed student has been declared dead. The "Every 15 Minutes" refers to the fact that someone in the U.S. dies every 15 minutes from alcohol-related traffic collision. The student 'victim' then returns to class with ghoulish makeup, a coroner's tag and a black Every 15 Minutes t-shirt. Since this victim has been declared dead, the student is not allowed to speak or interact with classmates for the rest of the day. The program can also include a simulated traffic collision and a mock funeral at the school, as well as trips to a hospital emergency room, jail, and the morgue. Advocates of the program claim that the program offers real-life experience without the real-life risks. The staged event is designed to instill in teenagers the potentially dangerous consequences of drinking alcohol in a dramatic fashion (see www.every15minutes.com).

Evaluation of the effectiveness of the Every 15 Minutes program has been mixed. Hover et al. (2000) compared two neighboring high schools in Springfield, Missouri, one of which presented the program and one that did not. The authors found that while the program did change some alcohol-related attitudes in the participants, alcohol-related behaviors did not change. By contrast, Bordin et al. (2003) evaluated the same program in 81 California high schools using 1651 students. Using a pre-post comparison of only the participants who played the living dead character (with no control group), the authors found that those students who played an active role in the presentation reported drinking less, being more likely to talk to their friends about drinking and driving, and were less likely to drive after drinking or ride with someone who had been drinking. Behaviors of observers (students who were part of the program but did not 'die') were not reported in this study. Further evaluation of the effectiveness of this program is warranted.

The widespread use of the "Every 15 Minutes" program became front page news in 2008 when things went seriously wrong at an El Camino High School located in Oceanside, California. The school had enlisted Highway Patrol officers to deceive students into believing that some of their classmates had actually died in car wrecks over the prior weekend. Since the students were not informed of this 'mock' exercise, the student body was seriously traumatized and some students were described as hysterical (Las Vegas Sun editorial, 2008). Thus far, use of the "Every 15 Minutes" program in its original form still remains quite popular, particularly in some states like California.

APPROACHES BY COLLEGES AND UNIVERSITIES

U.S. college administrators are acutely aware of the problem of binge drinking on their campuses. Many colleges have launched alcohol prevention and intervention programs for college students (Walters et al., 2000). One survey of 734 U.S. college administrators in 1999 indicated that prevention practices were widespread in the general areas of education about alcohol, use of policy controls to limit access to alcohol, restricting advertising of alcohol at home-game sporting events, and implementation of alcohol-free residences. Most of the surveyed colleges had a campus alcohol specialist, many had task forces, and about half of colleges were performing in-house data collection of drinking-related problems. Less prevalent on various campuses were targeted alcohol education, outreach, and restrictions on alcohol advertising in campus media. Also less common at that time were program evaluations, community agreements, and neighborhood exchanges (Wechsler et al., 2000). Thus, by the end of the 1990s, the administrators of colleges and universities in the U.S. seemed to be aware of the problem of binge drinking and were making attempts to address the problem. By the year 2000, a review of the success of various existing programs being offered indicated that general educational and abstinence-based approaches showed the least efficacy. By contrast, programs that incorporated learning new skills, changing attitudes, or feedback-based interventions seemed to be moderately more successful (Walters et al., 2000). However, the alcohol culture on college campuses was and still is a significant hindrance in changing the behaviors of students.

As college drinking continued to remain high or even escalate, a national task force was convened in 2002 at the National Institute on Alcohol Abuse and Alcoholism to evaluate the state of the problem and make recommendations on how the U.S. could address it. The results from this Task Force of the National Advisory Council on Alcohol Abuse and Alcoholism (2002) can be found on the website entitled, *College Drinking: Changing the Culture*, found at www.collegedrinkingprevention.gov. This website makes recommendations for high school administrators and associated parents and students as well as college presidents and associated parents and students. First, the task force compiled all of the published studies on the known annual consequences of college drinking (as discussed in Chapter 3), such as accident rates, death rates, etc. Based on the extensive review of all the available studies on college student drinking at that time, the task force also identified three constituencies that need to be addressed to improve the culture of drinking on college campuses in the U.S. These three include: 1) the college student as an individual, 2) the student population as a whole and 3) the college and the surrounding community. Thus, the task force argued that the needs of each of these three constituencies require attention and assessment as a prevention program for a college campus is developed and implemented. In other words, one might develop a prevention program that targets only at-risk or alcohol-dependent students. By contrast, a university administration might also develop a campus-wide prevention program that targets the whole campus community including the neighborhood where the campus is located.

To date, there have been several attempts at prevention of binge drinking at the college level. One popular approach is to attempt to change social norms regarding drinking among college students. Studies have shown that students often have misconceptions about how their peers view drinking. Generally, students view their peers as being more permissive in their personal attitudes toward drinking than actually is the case. This finding has been reported not only at schools where drinking occurs in moderation, but also at schools where drinking heavily is commonplace. For example, a student may believe that her friends think it is just fine for her to get so drunk that she throws up. By contrast, if you ask the friends, they actually think that drinking to the point of vomiting is drinking too much. Similarly, students are often inaccurate about drinking norms. For example, students believe their peers are consuming alcohol more frequently and more heavily than is actually the case. As such, these exaggerated perceptions of one's peers may lead a student to feel more pressure or greater license to conform to these incorrect expectations of their peers, thus leading to heavier and more frequent binge drinking (Perkins, 2002).

These extensive misperceptions about drinking norms have also been documented recently in Canadian university students, even though Canadian students tend to drink in a more moderate fashion compared to their U.S. counterparts (Kuo et al., 2002). One survey of over 5000 university students from 11 Canadian universities found that Canadian students also overestimate the alcohol consumptions norms (both quantity and frequency), regardless of the actual drinking norm on campus. Similar to the findings with U.S. students, Canadian students' perception of their campus drinking norm was a stronger predictor of personal alcohol use than the actual campus norm. Interestingly, students who personally abstain from alcohol and drink lightly felt slightly alienated from campus life, indicating that these misperceptions may hurt all students, regardless if they binge drink or not (Perkins, 2007).

If students have misconceptions about what the drinking norms truly are on campus, an appropriate prevention program might be to disseminate information about actual drinking norms. This "social norms approach" has been used on many campuses. The actual norms about drinking are publicized in various ways including orientation week programs for new students, in student newspapers, on student websites, in lectures, on radio stations, and on campus posters and flyers. The success of this approach is in debate. Some studies have reported that campuses with social norms prevention programs have witnessed significant decreases in high-risk drinking (Perkins, 2002).

Social norms interventions have also been used to target those individuals who are most at-risk on campus. A three-year longitudinal study of a social norms intervention with student athletes found that the intervention substantially reduced misperceptions about alcohol norms among student-athlete peers. The negative consequences of alcohol misuse decreased by 30% and personal misuse of alcohol decreased by 50% or more among the student-athletes after program exposure, in comparison to a group that had minimal program exposure (Perkins and Craig, 2006). The results of this study seem to indicate that application of social norms interventions with a high-risk college subpopulation (student athletes) reduces excessive and dangerous drinking in this subgroup that is likely to binge drink.

While the results of some evaluations of social norms programs have been impressive, others have found that this approach, while an extremely popular prevention approach, really hasn't changed the drinking culture on college campuses. For example, Wechsler et al. (2003) evaluated the social norms marketing interventions while looking at their extensive data sets from the Harvard School of Public Health College Alcohol Studies (CAS) from 1997, 1999 and 2001. From these data sets, 37 colleges had implemented social norms marketing programs and 61 did not over this time period. The authors found no decreases on any measures of alcohol use at schools with social norms program, even when student exposure and length of program existence were taken into account. More problematic for social norms advocates is that *increases* in measure of monthly alcohol use and total volume of alcohol consumed were observed at the schools employing social norms programs. As such, this one study with a very large and comprehensive data set seems to indicate that these social norms programs, at least in their current form, are not reducing alcohol use among college students.

An alternative prevention approach has been to focus on individuals and not the whole campus. A group at the University of Washington, led by Dr. Alan Marlatt has approached the problem of college drinking by working with students (either individually or in small groups) to develop skills that can be used to avoid alcohol-related problems. Their skills-training program focuses on four skills or components. The first is training in blood-alcohol-level monitoring so that students have adequate knowledge about specific alcohol effects. The second is to develop coping skills to use in situations associated with dangerous drinking. The third is to modify expectations regarding alcohol use and alcohol effects. Finally, the fourth builds the general skills of stress-management and other life-management skills (Baer et al., 1994; Kivlahan et al., 1990).

This approach is obviously labor-intensive and expensive to administer. However, the approach does seem to change drinking habits of participants. For example, students who participated in an eight-week skills-training program decreased the number of drinks per week, the peak blood alcohol level reached per week, and the hours per week with a blood alcohol concentration exceeding .055g%, compared to before the program. This pre-to-post change in drinking habits were not observed in other students in their control groups who only participated in an assessment phase or only attended an alcohol education class that emphasized alcohol effects. Furthermore, the change in drinking habits in the participants who attended this program was pervasive. When followed up 12 months after the skills-training program, those individuals were still drinking less, a fairly impressive finding (Baer et al., 1994).

Both the social norms approach and the skills-training approach involve a fair amount of resources to change student drinking, resources to which not every school may have access. Luckily, a recently published study suggests that students will change their drinking if they get caught only once for violating school policies about drinking. White et al. (2008) evaluated the efficacy of brief interventions with students who were caught drinking and reprimanded for violating university rules regarding substance use. In their study, half of the students when caught drinking on campus were both reprimanded by the university and giving a personalized feedback intervention to help them reduce their drinking. The other half of the students was just reprimanded. Both groups reduced their drinking and alcohol-related problems from baseline at both the 2 month and 7 month time point after the infraction. There were no significant differences between the groups. It appeared that just getting caught works. As such, schools with strict policies about underage drinking, public intoxication, vandalism of property etc. should be able to reduce binge drinking. A study has yet to be conducted to determine whether variance in how schools penalize drinking infractions alters student drinking behavior on campus.

Specific cases of success notwithstanding, the overall trends of college binge drinking even during a period of increased prevention efforts have not indicated any positive change for the better in binge drinking rates. Data from the College Alcohol Study (CAS) surveys completed by the Harvard School of Public health over the years 1993, 1997, 1999 and 2001 found that there was very little change in overall binge drinking occurring at the individual college level, despite implementation of various prevention programs. In fact, binge drinking was increasing among students attending all-women's colleges, a pattern not seen in the earlier surveys. This pattern is disheartening considering that, towards the end of the survey years, more students now lived in substance-free housing and encountered college educational efforts and greater sanctions resulting from their alcohol use (Wechsler et al., 2002). In other words, colleges are making an effort but the effort they are making is not really doing much of anything, if at all.

If the current approaches by colleges and universities are not working, a survey of college administrators indicates where progress could still be made (Wechsler et al., 2004). Most schools in this survey (68% of administrators from the 4-year colleges in the U.S. responded) conducted targeted alcohol education and invested in institutional prevention efforts. Half of the schools conducted social norms campaigns. These campaigns were funded by government agencies (1 in 3 schools) or by the alcohol industry itself (1 in 5 schools). Less common was actual restriction of alcohol use on campus or at college events. Thus, the authors of this study suggested that colleges may want to refocus their prevention efforts on the demand or supply end of the equation. Less alcohol on a campus could result in less binge drinking.

CONCLUSION

Various prevention programs aimed at curbing binge drinking have been implemented at the elementary school, high school and college level. In elementary schools and high schools, the Drug Abuse Resistance Education (DARE) program is almost universally accepted as the prevention program of choice in schools in the U.S. Despite the widespread popularity of the program, there are several concerns about its use. First, its efficacy in reducing drug and alcohol use in high school students has not been demonstrated. Second, underage drinking and binge drinking are not of foremost importance in the DARE curriculum. While methamphetamine or heroin use is dangerous, most high schools students may never encounter these drugs. Instead, statistics show that many high school students drink, and many binge drink, with deleterious health consequences. Clearly, a program that specifically targets the problem of binge drinking in adolescents is warranted.

The culture of alcohol on college campuses is of significant concern. Prevention programs at the college level, such as social norms programs, have also had limited success. Skills training programs have fared better in evaluations but have not been as yet widely implemented, probably due to the cost and labor-intensiveness of these programs. Thus, new ideas and efforts on prevention and intervention with the problem of binge drinking in adolescents and young adults are still clearly needed.

Chapter 18

BINGE DRINKING IN YOUNG PEOPLE: WHERE DO WE GO FROM HERE?

Throughout this book, we have tried to provide abundant evidence that binge drinking is a behavior with undesirable outcomes for both young drinkers and individuals around them. While many people in the general public think of high school and college binge drinking as a rite of passage, there is now a growing realization among the scientific community of how problematic and abnormal this type of drinking really is. Unfortunately, research on the harms and hazards of binge drinking have been limited by debates of how binge drinking should be defined and how many drinks make up a binge. These debates have been problematic for solving the binge drinking problem for two reasons. First, these definitional debates have hampered research efforts that seek to understand why binge drinking is so common in high school and college students, the demographic groups who are most likely to engage in this behavior. Second, these debates have delayed efforts to prevent and intervene in this harmful drinking pattern.

By providing the most recent and comprehensive evidence, we have argued binge drinking is a pattern of alcohol consumption that can create a variety of harms (e.g., health, personal, social, and academic harms) and result in a variety of poor outcomes (e.g., increased risk of future alcohol dependence). However, we think that the saddest statistics associated with binge drinking are the death rates. In the United States, approximately 1,400 college students between the ages of 18 and 24 die each year from alcohol-related incidents (NIAAA, 2002). Worldwide, 1.8 million deaths annually are the result of injuries caused by hazardous and harmful drinking. This accounts for 3.2% of all deaths and accounts for 4.0% of disease burden (WHO, 2007). However, the numbers do not do justice to the stories of these young individuals who showed great promise and died needlessly.

As of this writing, the most recent death appears to be that of an undergraduate student from Wabash College, an Indiana college. On October 6, 2008, an 18 year old student, Johnny Smith, was found dead on this small college campus. Smith was an aspiring engineer. The name and college may be new, but the story is familiar. Wabash College had beaten Allegheny College in a football game on Saturday night. Students celebrated with heavy drinking, including those students who were members of the Delta Tau Delta fraternity. Detectives informed Smith's family that four fraternity brothers had brought Smith back to his room very drunk on Saturday night. A fellow fraternity brother found Smith dead in a bed at the fraternity house on Sunday morning. Toxicology reports were still pending at the time of this writing, but it appears that binge drinking in this underage student is probably the cause of death. Wabash College President Pat White had to make that dreaded phone call. He called Smith's parents in Arizona to tell them the tragic news (Gillers et al., 2008; Rodriguez, 2008). While one wonders how a college president can tactfully make that phone call, unbelievable that was not the first phone call of that type President White has had to make this year. Another 18 year old freshman, Patrick Woehnker, died after falling off a roof of a campus building in October of 2007. The Coroner reported that alcohol was a contributing factor to Woehnker's death (Gillers et al., 2008; Rodriguez, 2008). Stories such as these highlight the completely preventable aspects of these deaths. If drinking was at a more moderate level, or not at all (in compliance with underage drinking laws), these deaths might not have occurred.

It seems that the brightest and the best of our young people are not immune to binge drinking and all of its associated problems. For example, American Olympic swimmer Michael Phelps captured everyone's heart and imagination in the Beijing Summer Olympics in 2008. This 23 year old now holds the record for the most gold medals won at a single Olympics, with the eight golds he won in Beijing. His performance was also impressive at the previous Summer Olympics in Athens in 2004. In Athens, he won six gold and two bronze medals. His swimming abilities are awe inspiring. His underage drinking and impaired driving while attending the University of Michigan between the 2004 and 2008 Olympics are not. In November 2004, Phelps, age 19, was stopped by a state trooper for running a stop sign in Maryland. The trooper detected a strong odor of alcohol when he approached the car and noted that Phelps' eyes were bloodshot and glassy. Phelps initially denied drinking, but then recanted when he failed road-side field sobriety tests. He was subsequently arrested for driving under the influence of alcohol. He pleaded guilty to driving while impaired and was granted probation before judgment, which meant that his record would be expunged if he complied with the terms of probation. He was ordered to serve 18 months probation, fined $250, obligated to speak to high school students about drinking and driving and had to attend a Mothers Against Drunk Driving (MADD) meeting. When questioned about the incident that month by Matt Lauer on the Today Show, Phelps called the arrest an "isolated incident". He also said that he had "definitely let myself down and my family down... I think I let a lot of people in the country down" (Associated Press, 2004; Michaelis, 2004).

While Phelps may have called his underage drinking and impaired driving an "isolated incident", he belongs to a large group of young people who drink irresponsibly and engage in harmful behaviors. What makes the Phelps case different from many others is that Phelps was extremely lucky that he did not hurt himself or anyone else with his actions. The World Health Organization (WHO) recently reported that young people in many countries around the world are beginning to drink at earlier ages, a pattern that is associated with greater likelihood of both alcohol dependence and alcohol-related injury later in life. This worldwide problem of underage binge drinking appears to be increasing in many countries, despite educational approaches on the prevention of alcohol problems among young people (WHO, 2001).

In the year 2000, the U.S. Surgeon General established a 50% reduction in college binge drinking by the year 2010 as one of the health goals for the United States (U.S. Department of Health and Human Services, 2000). The Centers for Disease Control and Prevention provides measures of binge drinking in their state-by-state reports to assess the outcome of such a goal (Division of Adult and Community Health, National Center for Chronic Disease Prevention and Health Promotion, Centers for Disease Control and Prevention, 1995-1999). Binge drinking has been on the national radar screen and clearly has become a priority for many people. Both the U.S. Senate and the House of Representatives adopted resolutions, introduced by Senator Joseph Biden (Senate Resolution 192, 1998) and Representative Joseph Kennedy (House Resolution 321, 1997) calling for national action to address college binge drinking. The World Health Organization stressed the importance of addressing the binge drinking problem by developing a conference on the topic (World Health Organization, 2001). Despite all of these admirable efforts to attempt to track rates of binge drinking and efforts to reduce this behavior, there appears to very limited success in changing this behavior. As of this writing in 2008, the United States is nowhere near the original goal of a 50% reduction in binge drinking from 2000 levels, and in fact, levels of binge drinking in young people in the United States appear to be relatively unchanged (Fournier and Levy, 2006; SAMHSA, 2006).

The constancy of binge drinking behavior in young people, despite increased attention to the problem, begs the question of what novel approaches might lead to change. In our review of current approaches to prevention and intervention in Chapter 17, we were surprised at the general ineffectiveness of current approaches. In elementary and high schools, the Drug Abuse Resistance Education (DARE) program is almost universally accepted as the prevention program of choice in schools in the U.S. Despite the widespread popularity of the program, there are two significant concerns. First, binge drinking is not of foremost concern in the DARE curriculum since it is a general drug resistance program. While prevention of use of all drugs is important, alcohol is the primary drug used by underage people. Alcohol needs special focus. Second, the efficacy of DARE in reducing drug and alcohol use in high school students has not been clearly demonstrated. Similar concerns exist for prevention programs at the college level, such as social norms programs, that have also had limited success. Skills training programs have fared somewhat better in evaluations, but have not yet been widely implemented, probably due to the cost and labor-intensiveness of these programs.

Despite the limited effectiveness of widespread prevention programs, more successful outcomes have been seen in intervention programs. For example, adolescents and college students who present to the emergency department are potentially in an enhanced state of receptiveness to intervention and readiness to change their drinking behavior (see Chapter 10). Studies have demonstrated that a screening and brief intervention that occurs right at the time of presentation reduces future hazardous drinking and its associated risk of injury, illness, and even death. While the primary prevention of alcohol related injury or illness is clearly the ideal, these types of interventions appear helpful and should be more universally adopted.

Wechsler and Wuethrich (2002) had several other suggestions of what young people, parents and schools can change to reduce binge drinking. They argued that good communication between young people and their parents is paramount. Asking adolescents to engage in self-control and resist pressure from friends to drink needs to be a topic of conversation well before these specific issues come to fruition. Likewise, setting rules and enforcing consequences for drinking should be well established. In this book, we have provided ample evidence that underage binge drinking may lead to brain damage and other risks to health and well-being. As such, enabling underage drinking by providing alcohol to children or allowing any sort of underage consumption should be considered unacceptable.

Likewise, schools (elementary school, high school and college level) need to have strict enforcement of violations of alcohol policies. Particularly at the college level, alcohol control policies are often loosely enforced if at all. However, simply enforcing a multitude of rules is not enough. People need to know why these rules are important. An excellent example is changes in methamphetamine use in the U.S. Initially, use of crystal meth was increasing. However, when the general public became acutely aware of the significant health damage caused by the use of methamphetamine by public health campaigns and television programs, use decreased in both adolescents and other high-risk groups (Fournier and Levy, 2006; Reback et al., 2008). Good communication about the health effects of binge drinking might similarly reduce this pattern of alcohol consumption. In addition, good communication within families may not be that difficult to establish. Simply having dinner together as a family has been demonstrated as a variable that is associated with later alcohol use initiation in teenage girls (see Chapter 14).

Children model the behavior of adults around them. If those adults model the responsible use of alcohol, their children are also more likely to do so. In this book, we described studies that demonstrate that children have very detailed expectancies about alcohol and its use from a very early age (see Chapter 15). Parents who think that their young children are unaware of their alcohol consumption patterns are deceiving themselves. Parents also need to be aware of risk factors for binge drinking and be particularly vigilant when those risk factors are present. For example, children who have their first drink at a younger age are more likely to binge drink and develop an alcohol dependence problem. Personality traits such as impulsivity and sensation seeking are associated with binge drinking. Finally, parents and schools need to know that friends matter a great deal. In high school, if a child has many friends who drink, that child is likely to drink. In college students, those who are members of fraternities, sororities, or college athletic organizations are more likely to binge drink. The choice of school matters as well. Many high school students who drank little became binge drinkers once they became college students at schools with high rates of binge drinking (Wechsler and Wuethrich, 2002).

In conclusion, there are a multitude of factors that contribute to the high rates of binge drinking in our young people. This is not a simple problem that can be solved by fixing just one variable. However, one thing that has become clear is that ignoring this problem will not change the drinking behavior of our adolescents and college students. The scientific evidence is accumulating that this pattern of alcohol consumption is harmful, particularly for a brain that is still developing. Thus, novel approaches to limit this type of excessive drinking behavior are needed. While not all efforts will succeed, we need new ideas that go beyond our current approaches. Most of all, we need to counteract this common myth that binge drinking is a rite of passage, a fun way to transition from childhood to adult responsibilities. This myth contributes to injuries and deaths and needs to be challenged by whatever means necessary.

ACKNOWLEDGMENTS

We gratefully acknowledge a number of individuals who made this book possible:

Dr. Mark Fillmore at the University of Kentucky for his insightful advice over many years and his steadfast support and encouragement. His detailed knowledge about alcohol use and abuse is unparalleled in the field and his research has changed how we think about acute effects of alcohol in social drinkers.

The Alcoholic Beverage Medical Research Foundation for their grant support and recognition of the vital importance of supporting new investigators in the field of alcohol research.

Cecile Marczinski would like to thank her husband Chris Horn and her daughter Isabella for their love and support. Chris provided sage advice through-out this process and he understands that thinking follows language. Our conversations about binge drinking have provided the framework for my research and my writing.

Estee and Vincent Grant would like to thank their families for all of their love, support and guidance over the years. They would also like to thank all health care workers who provide care to these vulnerable adolescents and young adults, as well as all those working on alcohol and injury prevention programs.

Nova Science Publishers, Inc., for their enthusiasm and support for this project.

REFERENCES

Aasebo, W., Erikssen, J., Jonsbu, J., and Stavem, K. (2007). ECG changes in patients with acute ethanol intoxication. *Scandinavian Cardiology Journal, 41*, 79-84.

Abroms, B.D., Fillmore, M.T., and Marczinski, C.A. (2003). Alcohol-induced impairment of behavioral control: Effects on the alteration and suppression of prepotent responses. *Journal of Studies on Alcohol, 64*, 687-695.

Academic ED SBIRT Research Collaborative (2007). The impact of screening, brief intervention, and referral for treatment on emergency department patients' alcohol use. *Annals of Emergency Medicine, 50*, 699-710.

Acton, G.S. (2003). Measurement of impulsivity in a hierarchical model of personality traits: Implications for substance use. *Substance Use and Misuse, 38*, 67-83.

Adesso, V.J. (1985). Cognitive factors in alcohol and drug use. In M. Galizio and S.A. Maisto (Eds.), *Determinants of substance use* (pp. 179-208). New York, NY: Plenum Press.

Agartz, I., Momenan, R., Rawlings, R.R., Kerich, M.J., and Hommer, D.W. (1999). Hippocampal volume in patients with alcohol dependence. *Archives of General Psychiatry, 56*, 356-363.

Alcoholics Anonymous (1972). *If you are a professional, A.A. wants to work with you.* New York, NY: Alcoholics Anonymous World Services.

Alcoholics Anonymous (1983). *Questions and answers on sponsorship.* New York, NY: Alcoholics Anonymous World Services.

Alford, C., Cox, H., and Wescott, R. (2001). The effects of red bull energy drink on human performance and mood. *Amino Acids, 21*, 139-150.

Allely, P., Graham, W., McDonnell, M., and Spedding, R. (2006). Alcohol levels in the emergency department: a worrying trend. *Emergency Medicine Journal, 23*, 707-708.

American Medical Association (2003). *Proceedings from the Educational Forum on Adolescent Health: Youth Drinking Patterns and Alcohol Advertising.* Chicago, IL: Author.

American Psychiatric Association (1987). *Diagnostic and Statistical Manual of Mental Disorders* (3rd ed. rev.). Washington, DC: Author.

American Psychiatric Association (1994). *Diagnostic and Statistical Manual of Mental Disorders* (4th ed.). Washington, DC: Author.

American Psychiatric Association (2000). *Diagnostic and Statistical Manual of Mental Disorders* (4th ed., rev.). Washington, DC: Author.

Anderson, S.W., Bechara, A., Damasio, H., Tranel, D., and Damasio, A.R. (1999). Impairment of social and moral behavior related to early damage in human prefrontal cortex. *Nature Neuroscience, 2*, 1032-1037.

Anderson, J.A., McLellan, B.A., and Pagliarello, G. (1990). The relative influence of alcohol and seatbelt usage on severity of injury from motor vehicle crashes. *Journal of Trauma, 30*, 415-417.

Arif, A.A., and Rohrer, J.E. (2005). Patterns of alcohol drinking and its association with obesity: data from the Third National Health and Nutrition Examination Survey, 1988-1994. *BMC Public Health, 5*, 126.

Arruda, E.P., Gomez, M., and Banfield, J. (2007). Effectiveness of the P.A.R.T.Y. (Prevent Alcohol and Risk-Related Trauma in Youth) program in changing youth driving-related attitudes. Presentation at the 2007 Canadian Injury Prevention and Safety Promotion Conference in Toronto, ON, Canada.

Associated Press (2004). Olympic champ sentenced for DUI. *CBS News,* retrieved October 11, 2008, from http://www.cbsnews.com/stories/2004/11/08/national.

Austin, E.W., and Johnson, K.K. (1997). Effects of general and alcohol-specific media literacy training on children's decision making about alcohol. *Journal of Health Communication, 2*, 17-42.

Baer, J.S., Kivlahan, D.R., Blume, A.W., McKnight, P., and Marlatt, G.A. (2001). Brief intervention for heavy drinking college students: 4-year follow-up and natural history. *American Journal of Public Health, 91*, 1310-1316.

Baer, J.S., Kivlahan, D.R., Fromme, K., and Marlatt, G.A. (1994). Secondary prevention of alcohol abuse with college student populations: A skills-training approach. In G. Howard and P.E. Nathan (Eds.), *Alcohol use and misuse by young adults* (pp. 83-108). Notre Dame, IN: Notre Dame University Press.

Bahr, S.J., Hoffmann, J.P., and Yang, X. (2005). Parental and peer influences on the risk of adolescent drug use. *Journal of Primary Prevention, 26*, 529-551.

Baird, J., Longabaugh, R., Lee, C.S., Nirenberg, T.D., Woolard, R., Mello, M.J., Becker, B., Carty, K., Allison Minugh, P., Stein, L., Clifford, P.R., and Gogineni, A. (2007). Treatment completion in a brief motivational intervention in the emergency department: the effect of multiple interventions and therapists' behavior. *Alcoholism: Clinical and Experimental Research, 31(10 Suppl)*, 71s-75s.

Barkley, R.A., Fischer, M., Smallish, L., and Fletcher, K. (2002). The persistence of attention-deficit/hyperactivity disorder into young adulthood as a function of reporting source and definition of disorder. *Journal of Abnormal Psychology, 111*, 279-289.

Barkley, R.A., Fischer, M., Smallish, L., and Fletcher, K. (2006). Young adult outcome of hyperactive children: Adaptive functioning in major life activities. *Journal of the American Academy of Child and Adolescent Psychiatry, 45*, 192-202.

Barnett, N.P., Goldestein, A.L., Murphy, J.G., Colby, S.M., and Monti, P.M. (2006). "I'll never drink like that again": characteristics of alcohol-related incidents and predictors of motivation to change in college students. *Journal of Studies on Alcohol, 67*, 754-763.

Barnett, N.P., Lebeau-Craven, R., O'Leary, T.A., Colby, S.M., Rohsenow, D.J., Monti, P.M., Wollard, R., and Spirito, A. (2002). Predictors of motivation to change after medical treatment for drinking-related events in adolescents. *Psychology of Addictive Behaviors, 16*, 106-112.

Barnett, N.P., Monti, P.M., Spirito, A., Colby, S.M., Rohsenow, D.J., Ruffolo, L., and Wollard, R. (2003). Alcohol use and related harm among older adolescents treated in an emergency department: the importance of alcohol status and college status. *Journal of Studies on Alcohol, 64*, 342-349.

Barnett, N.P., Spirito, A., Colby, S.M., Vallee, J.A., Woolard, R., Lewander, W., and Monti, P.M. (1998). Detection of alcohol use in adolescent patients in the emergency department. *Academic Emergency Medicine, 5*, 607-612.

Bartholow, B.D., Sher, K.J., and Krull, J.L. (2003). Changes in heavy drinking over the third decade of life as a function of collegiate fraternity and sorority involvement: a prospective, multilevel analysis. *Health Psychology, 22*, 616-626.

Baune, B.T., Mikolajczyk, R.T., Reymann, G., Duesterhaus, A., Fleck, S., Kratz, H., and Sundermann, U. (2005). A 6-months assessment of the alcohol-related clinical burden at emergency rooms (ERs) in 11 acute care hospitals of an urban area in Germany. *BMC Health Services Research, 5*, 73.

Bazargan-Hejazi, S., Bazargan, M., Gaines, T., and Jemanez, M. (2008). Alcohol misuse and report of recent depressive symptoms among ED patients. *American Journal of Emergency Medicine, 26*, 537-544.

Bazargan-Hejazi, S., Bing, E., Bazargan, M., Der-Martirosian, C., Hardin, E., Bernstein, J., and Bernstein, E. (2005). Evaluation of a brief intervention in an inner-city emergency department. *Annals of Emergency Medicine, 46*, 67-76.

Bazargan-Hejazi, S., Gaines, T., Duan, N., and Cherpitel, C.J. (2007). Correlates of injury among ED visits: effects of alcohol, risk perception, impulsivity, and sensation seeking behaviors. *American Journal of Drug and Alcohol Abuse, 33*, 101-108.

Bechara, A., Dolan, S., Denburg, N., Hindes, A., Anderson, S.W., and Nathan, P.E. (2001). Decision-making deficits, linked to a dysfunctional ventromedial prefrontal cortex, revealed in alcohol and stimulant abusers. *Neuropsychologica, 39*, 376-389.

Beirness, D.J. (1987). Self-estimates of blood alcohol concentration in drinking-driving context. *Drug and Alcohol Dependence, 19*, 79-90.

Beirness, D.J., Foss, R.D., and Vogel-Sprott, M. (2004). Drinking on campus: Self-reports and breath tests. *Journal of Studies on Alcohol, 65*, 600-604.

Beirness, D.J., and Vogel-Sprott, M. (1984). The development of alcohol tolerance: Acute recovery as a predictor. *Psychopharmacology, 84*, 398-401.

Bennett, R.H., Cherek, D.R., and Spiga, R. (1993). Acute and chronic alcohol tolerance in humans: Effects of dose and consecutive days of exposure. *Alcoholism: Clinical and Experimental Research, 17*, 740-745.

Bernadt, M.W. (1982). Comparison of questionnaire and laboratory tests in the detection of excessive drinking and alcoholism. *Lancet, 6*, 325–328.

Bernstein, E., Bernstein, J., and Levenson, S. (1996). Project ASSERT: and ED-based intervention to increase access to primary care, preventive services, and the substance abuse treatment system. *Annals of Emergency Medicine, 30*, 181-189.

Beseler, C.L., Aharonovich, E., Keyes, K.M., and Hasin, D.S. (2008). Adult transition from at-risk drinking to alcohol dependence: the relationship of family history and drinking motives. *Alcoholism: Clinical and Experimental Research, 32*, 607-616.

Bichler, G., and Tibbetts, S.G. (2003). Conditional covariation of binge drinking with predictors of college students' cheating. *Psychological Reports, 93*, 735-749.

Biyik, I., and Ergene, O. (2006). Acute myocardial infarction associated with heavy alcohol intake in an adolescent with normal coronary arteries. *Cardiology in the Young, 16*, 190-192.

Bjork, J., Hommer, D.W., Grant, S.J., and Danube, C. (2004). Impulsivity in abstinent alcohol-dependent patients: relation to control subjects and type 1-/type 2-like traits. *Alcohol, 34*, 133-150.

Blow, F.C., Barry, K.L., Walton, M.A., Maio, R.F., Chermack, S.T., Bingham, C.R., Ignacio, R.V., and Strecher, V.J. (2006) The efficacy of two brief intervention strategies among injured, at-risk drinkers in the emergency department: impact of tailored messaging and brief advice. *Journal of Studies on Alcohol, 67*, 568-578.

Bordin, J., Bumpus, M., and Hunt, S. (2003). Every 15 Minutes: A preliminary evaluation of a school based drinking/driving prevention program. *Californian Journal of Health Promotion, 1*, 1-6.

Borges, G., Cherpitel, C., and Mittleman, M. (2004). Risk of injury after alcohol consumption: a case-crossover study in the emergency department. *Social Science Medicine, 58*, 1191-1200.

Borges, N.J., and Hansen, S.L. (1993). Correlation between college students' driving offenses and their risks for alcohol problems. *Journal of American College Health, 42*, 79-81.

Borlikova, G.G., Elbers, N.A., and Stephens, D.N. (2006). Repeated withdrawal from ethanol spares contextual fear conditioning and spatial learning but impairs negative patterning and induces over-responding: evidence for effect on frontal cortical but not hippocampal function? *European Journal of Neuroscience, 24*, 205-216.

Botvin, G.J., and Griffin, K.W. (2003). Drug abuse prevention curricula in schools. In Z. Sloboda and W.J. Bukoski (Eds.), *Handbook of drug abuse prevention: Theory, science, and practice* (pp. 45-74). New York: Kluwer Academic/Plenum Publishers.

Boyatzis, R.E. (1975). The predisposition toward alcohol-related interpersonal aggression in men. *Journal of Studies on Alcohol, 36*, 1196-1207.

Bradley, K.A., Donovan, D.M., and Larson, E.B. (1993). How much is too much? Advising patients about safe levels of alcohol consumption. *Archives of Internal Medicine, 153*, 2734-2740.

Brenner, J., and Swanik, K. (2007). High-risk drinking characteristics in collegiate athletes. *Journal of American College Health, 56*, 267-272.

Brown, S.A. (1985). Expectancies versus background in the prediction of college drinking patterns. *Journal of Consulting and Clinical Psychology, 53*, 123-130.

Brown, S.A., Goldman, M.S., and Christiansen, B.A. (1985). Do alcohol expectancies mediate drinking patterns of adults? *Journal of Consulting and Clinical Psychology, 53*, 512-519.

Brown, S.A., Tapert, S.F., Granholm, E., and Delis, D.C. (2000). Neurocognitive functioning of adolescents: effects of protracted alcohol use. *Alcoholism: Clinical and Experimental Research, 24* 164-171.

Bruce, G., and Jones, B.T. (2004). A pictorial Stroop paradigm reveals an alcohol attentional bias in heavier compared to lighter social drinkers. *Journal of Psychopharmacology, 18*, 527-533.

Brumback, T., Cao, D., and King, A. (2007). Effects of alcohol on psychomotor performance and perceived impairment in heavy binge social drinkers. *Drug and Alcohol Dependence, 91*, 10-17.

Bushman, B.J., and Cooper, H.M. (1990). Effects f alcohol on human aggression: An integrative research review. *Psychological Bulletin, 107*, 341-354.

Bussey, T.J., Dias, R., Redhead, E.S., Pearce, J.M., Muir, J.L., and Aggleton, J.P. (2000). Intact negative patterning in rats with fornix or combined perirhinal and postrhinal cortex lesions. *Experimental Brain Research, 134*, 506-519.

Cain, M.E., Saucier, D.A., and Bardo, M.T. (2005). Novelty seeking and drug use: Contribution of an animal model. *Experimental and Clinical Psychopharmacology, 13*, 367-375.

Cameron, C.A., Stritzke, W.G., and Durkin, K. (2003). Alcohol expectancies in late childhood: an ambivalence perspective on transitions toward alcohol use. *Journal of Child Psychology and Psychiatry, 44*, 687-698.

Capone, C., and Wood, M.D. (2008). Density of familial alcoholism and its effects on alcohol use and problems in college students. *Alcoholism: Clinical and Experimental Research, 32*, 1451-1458.

Capone, C., Wood, M.D., Borsari, B., and Laird, R.D. (2007). Fraternity and sorority involvement, social influences, and alcohol use among college students: a prospective examination. *Psychology of Addictive Behaviors, 21*, 316-327.

Carlini-Marlatt, B., Gazal-Carvalho, C., Gouveia, N., and Souza Mde F. (2003). Drinking practices and other health-related behaviors among adolescents of Sao Paulo City, Brazil. *Substance Use and Misuse, 38*, 905-932.

Casey, B.J., Giedd, J.N., and Thomas, K.M. (2000). Structural and functional brain development and its relation to cognitive development. *Biological Psychology, 54*, 241-257.

Cashin, J.R., Presley, C.A., and Meilman, P.W. (1998). Alcohol use in the Greek system: follow the leader? *Journal of Studies on Alcohol, 59*, 63-70.

Cato, M.A., Delis, D.C., Abildskov, T.J., and Bigler, E. (2004). Assessing the elusive cognitive deficits associated with ventromedial prefrontal damage: a case of a modern-day Phineas Gage. *Journal of the International Neuropsychological Society, 10*, 453-465.

Chamberlain, E., and Solomon, R. (2002). The case for a 0.05% criminal law blood-alcohol concentration limit for driving. *Injury Prevention, 8 (Suppl. III)*, iii1-iii17.

Chambers, R.A., Taylor, J.R., and Potenza, M.N. (2003). Developmental neurocircuitry of motivation in adolescence: A critical period of addiction vulnerability. *American Journal of Psychiatry, 160*, 1041-1052.

Champion, H.L., Foley, K.L., DuRant, R.H., Hensberry, R., Altman, D., and Wolfson, M. (2004). Adolescent sexual victimization, use of alcohol and other substances, and other health risk behaviors. *Journal of Adolescent Health, 35*, 321-328.

Chenet, L., Britton, A., Kalediene, R., and Petrauskiene, J. (2001). Daily variations in deaths in Lithuania: the possible contribution of binge drinking. *International Journal of Epidemiology, 30*, 743-748.

Cherpitel, C.J. (1993). Alcohol, injury, and risk-taking behavior: data from a national sample. *Alcohol: Clinical and Experimental Research, 17*, 762-766.

Cherpitel, C.J. (1996). Drinking patterns and problems and drinking in the event: an analysis of injury by cause among casualty patients. *Alcohol: Clinical and Experimental Research, 20*, 1130-1137.

Cherpitel, C.J., Ye, Y., and Bond, J. (2004). Alcohol and injury: multi-level analysis from the emergency room collaborative alcohol analysis project (ERCAAP). *Alcohol and Alcoholism, 39*, 552-558.

Cherpitel, C.J., Ye, Y., Bond, J., Rehm, J., Cremonte, M., Neves, O., Moskalewicz, J., Swiatkiewicz, G., and Giesbrecht, N. (2006). The effect of alcohol consumption on emergency department services use among injured patients: A cross-national emergency room study. *Journal of Studies on Alcohol, 67*, 890-897.

Cherpitel, C.J., Ye, Y., Bond, J., Rehm, J., Poznyak, V., Macdonald, S., Stafstrom, M., and Hao, W. (2005). Multi-level analysis of alcohol-related injury among emergency department patients: a cross-national study. *Addiction, 100*, 1840-1850.

Choi, S., and Kellogg, C.K. (1992). Norepinephrine utilization in the hypothalamus of the male rat during adolescent development. *Developmental Neuroscience, 14*, 369-376.

Choi, D.S., Ward, S.J., Messaddeq, N., Launay, J.M., and Maroteaux, L. (1997). 5-HT receptors-mediated serotonin morphogenetic functions in mouse cranial neural crest and myocardiac cells. *Development, 124*, 1745-1755.

Chou, S.P., and Pickering, R.P. (1992). Early onset of drinking as a risk factor for lifetime alcohol-related problems. *British Journal of Addiction, 87*, 1199-1204.

Christiansen, B.A., and Goldman, M.S. (1983). Alcohol-related expectancies versus demographic/background variables in the prediction of adolescent drinking. *Journal of Consulting and Clinical Psychology, 51*, 249-257.

Christiansen, B.A., Goldman, M.S., and Inn, A. (1982). The development of alcohol-related expectancies in adolescents: Separating pharmacological from social learning influences. *Journal of Consulting and Clinical Psychology, 50*, 336-344.

Christiansen, B.A., Smith, G.T., Roehling, P.V., and Goldman, M.S. (1989). Using alcohol expectancies to predict adolescent drinking after one year. *Journal of Consulting and Clinical Psychology, 57*, 93-99.

Chung, T., Colby, S.M., Barnett, N.P., Rohsenow, D.J., Spirito, A., and Monti, P.M. (2000). Screening adolescents for problem drinking: performance of brief screens against DSM-IV alcohol diagnoses. *Journal of Studies on Alcohol, 61*, 579-587.

Chutuape, M.D., Mitchell, S.H., and de Wit, H. (1994). Ethanol preloads increase preference under concurrent random-ratio schedules in social drinkers. *Experimental and Clinical Psychopharmacology, 2*, 310-318.

Cicero, T.J. (1980). Alcohol self-administration, tolerance, and withdrawal in humans and animals: Theoretical and methodological issues. In H.Rigter and J.Crabbe Jr. (Eds), *Alcohol Tolerance and Dependence* (pp. 1-50). Amsterdam: Elsevier/North-Holland Biomedical Press.

Clark, D.B., Thatcher, D.L., and Tapert, S.F. (2008). Alcohol, psychological dysregulaton, and adolescent brain development. *Alcoholism: Clinical and Experimental Research, 32*, 375-385.

Cloninger, C.R. (1987). Recent advances in family studies of alcoholism. *Progress in Clinical and Biological Research, 241*, 47-60.

Cloninger, C.R., Sigvardsson, S., and Bohman, M. (1988). Childhood personality predicts alcohol abuse in young adults. *Alcoholism: Clinical and Experimental Research, 12*, 494-505.

Colby, S.M., Barnett, N.P., Eaton, C.A., Spirito, A., Woolard, R., Lewander, W., Rohsenow, D.J., and Monti, P.M. (2002). Potential biases in case detection of alcohol involvement among adolescents in an emergency department. *Pediatric Emergency Care, 18*, 350-354.

Collins, J.J., Jr. (Ed.). (1980). *Alcohol Use and Criminal Behavior: An Empirical, Theoretical, and Methodological Overview.* New York, NY: Guilford Press.

Connors, G.J., O'Farrell, T.J., Cutter, H.S.G., and Thompson, D.L. (1986). Alcohol expectancies among male alcoholics, problem drinkers, and nonproblem drinkers. *Alcoholism: Clinical and Experimental Research, 10*, 667-671.

Cooper, M.L., Russell, M., and George, W.H. (1988). Coping, expectancies, and alcohol abuse: A test of social learning formulations. *Journal of Abnormal Psychology, 97*, 218-230.

Corder, R., Mullen, W., Khan, N.Q., Marks, S.C., Wood, E.G., Carrier, M.J., and Crozier, A. (2006). Oenology: red wine procyanidins and vascular health. *Nature, 444*, 566.

Cranford, J.A., McCabe, S.E., and Boyd, C.J. (2006). A new measure of binge drinking: prevalence and correlates in a probability sample of undergraduates. *Alcoholism: Clinical and Experimental Research, 30*, 1896-1905.

Crews, F.T. (1999). Alcohol and neurodegeneration. *CNS Drug Review, 5*, 379-394.

Crews, F.T., Braun, C.J., Hoplight, B., Switzer, R.C. III, and Knapp, D.J. (2000). Binge ethanol consumption causes differential brain damage in young adolescent rats compared with adult rats. *Alcoholism: Clinical and Experimental Research, 24*, 1712-1723.

Crews, F., He, J., and Hodge, C. (2007). Adolescent cortical development: a critical period of vulnerability for addiction. *Pharmacology, Biochemistry and Behavior, 86*, 189-199.

Cruz, I.Y., and Dunn, M.E. (2003). Lowering risk for early alcohol use by challenging alcohol expectancies in elementary school children. *Journal of Consulting and Clinical Psychology, 71*, 493-503.

Cunningham, R.M., Maio, R.F., Hill, E.M., and Zink, B.J. (2002). The effects of alcohol on head injury in the motor vehicle crash victim. *Alcohol and Alcoholism, 37*, 236-240.

Daeppen, J.B., Gaume, J., Bady, P., Yersin, B., Calmes, J.M., Givel, J.C., and Gmel, G. (2007). Brief alcohol intervention and alcohol assessment do not influence alcohol use in injured patients treated in the emergency department: a randomized controlled clinical trial. *Addiction, 102*, 1224-1233.

Dahchour, A., and De Witte, P. (1999). Effect of repeated ethanol withdrawal on glutamate microdialysate n the hippocampus. *Alcoholism: Clinical and Experimental Research, 23*, 1698-1703.

Damasio, A. (1999). *The Feeling of What Happens.* New York: Harcourt Brace.

Damasio, A., Grabowski, T., Frank, R., Galaburda, A.M., and Damasio, A.R. (1994). The return of Phineas Gage: clues about the brain from the skull of a famous patient. *Science, 264*, 1102-1105.

Darkes, J., and Goldman, M.S. (1993). Expectancy challenge and drinking reduction: Experimental evidence for a meditational process. *Journal of Consulting and Clinical Psychology, 61*, 344-353.

Davis, R. (2004). Five binge-drinking deaths 'just the tip of the iceberg'. *USA Today, October 7, 2004.*

Davis, R., and DeBarros, A. (2006). In college, first year is by far the riskiest. *USA Today*, January 24, 2006.

Dawson, D.A. (1998). Volume of ethanol consumption: effects of different approaches to measurement. *Journal of Studies on Alcohol, 59*, 191-197.

Deckel, A.W., Bauer, L., and Hesselbrock, V. (1995). Anterior brain dysfunctioning as a risk factor in alcoholic behaviors. *Addiction, 90*, 1323-1334.

de Groot, L.C., and Zock, P.L. (1998). Moderate alcohol intake and mortality. *Nutrition Review, 56*, 25-26.

Dent, A.W., Weiland, T.J., Phillips, G.A., and Lee, N.K. (2008). Opportunistic screening and clinician-delivered brief intervention for high-risk alcohol use among emergency department attendees: a randomized controlled trial. *Emergency Medicine Australasia, 20*, 121-128.

Department of Health, Home Office, Department for Education and Skills and Department for Culture, Media and Sport (2007). *Safe. Sensible. Social. The Next Steps in the National Alcohol Strategy.* Department of Health, London, UK.

Desousa, C., Murphy, S., Roberts, C., and Anderson, L. (2008). School policies and binge drinking behaviours of school-aged children in Wales – a multilevel analysis. *Health Education Research, 23*, 259-271.

Desapriya, E.B.R. (2004). Alcohol limit for drink driving should be much lower. *British Medical Journal, 328*, 895.

de Wit, H. (1996). Priming effects with drugs and other reinforcers. *Experimental and Clinical Psychopharmacology, 4*, 5-10.

De Wit, H., and Chutuape, M.D. (1993). Increased ethanol choice in social drinkers following ethanol preload. *Behavioral Pharmacology, 4*, 29-36.

de Wit, H., Crean, J., and Richards, J.B. (2000). Effects of *d*- amphetamine and ethanol on a measure of behavioral inhibition in humans. *Behavioral Neuroscience, 114*, 830-837.

de Wit, H., and Griffiths, R.R. (1991). Testing the abuse liability of anxiolytic and hypnotic drugs in humans. *Drug and Alcohol Dependence, 28*, 83-111.

Dick, D.M., Pagan, J.L., Holliday, C., Viken, R., Pulkkinen, L., Kaprio, J., and Rose, R.J. (2007). Gender differences in friends' influences on adolescent drinking: a genetic epidemiology study. *Alcoholism: Clinical and Experimental Research, 31*, 2012-2019.

Division and Adult and Community Health, National Center for Chronic Disease Prevention and Health Promotion, Centers for Disease Control and Prevention (1995-1999). *Behavioral Risk Factor Surveillance System online prevalence data.* Available at http://www.cdc.gov.

D'Onofrio, G., Becker, B., and Woolard, R.H. (2006). The impact of alcohol, tobacco, and other drug use and abuse in the emergency department. *Emergency Medicine Clinics of North America, 24*, 925-967.

D'Onofrio, G., Pantalon, M.V., Degutis, L.C., Fiellin, D.A., Busch, S.H., Chawarski, M.C., Owens, P.H., and O'Connor, P.G. (2008). Brief intervention for hazardous and harmful drinkers in the emergency department. *Annals of Emergency Medicine, 51*, 742-750.

D'Onofrio, G., Pantalon, M.V., Degutis, L.C., Fiellin, D.A., and O'Connor, P.G. (2005). Development and implementation of an emergency practitioner-performed brief intervention for hazardous and harmful drinkers in the emergency department. *Academy of Emergency Medicine, 12*, 249-256.

Drummond, C., Oyefeso, A., Phillips, T., Cheeta, S., Deluca, P., Winfield, H., Jenner, J., Cobain, C., Galea, S., Saunders, V., Fuller, T., Pappalardo, D., Baker, O., and

Christoupoulos, A. (2005). *Assessment Research Project (ANARP): The 2004 National Alcohol Needs Assessment for England.* Department of Health, London, UK.

Duka, T., Townshend, J.M., Collier, K., and Stephens, D.N. (2002). Kindling of withdrawal: a study of craving and anxiety after multiple detoxifications in alcoholic inpatients. *Alcoholism: Clinical and Experimental Research, 27,* 1563-1572.

Duka, T., Townshend, J.M., Collier, K., and Stephens, D.N. (2003). Impairment of cognitive functions after multiple detoxifications in alcoholic inpatients. *Alcoholism: Clinical and Experimental Research, 27,* 1563-1572.

Duka, T., Gentry, J., Malcolm, R., Ripley, T.L., Borlikova, G., Stephens, D.N., Veatch, L.M., Becker, H.C., and Crews, F.T. (2004). Consequences of multiple withdrawals from alcohol. *Alcoholism: Clinical and Experimental Research, 28,* 233-246.

Dumas, T.C., and Foster, T.C. (1998). Late developmental changes in the ability of adenosine A1 receptors to regulate synaptic transmission in the hippocampus. *Developmental Brain Research, 105,* 137-139.

Duncan, D.F. (1997). Chronic drinking, binge drinking, and drunk driving. *Psychological Reports, 80,* 681-682.

Duncan, J. (1981). Directing attention in the visual field. *Perception and Psychophysics, 30,* 90-93.

Dunn, M.E., and Goldman, M.S. (1998). Age and drinking-related differences in the memory organization of alcohol expectancies in 3rd-, 6th-, 9th-, and 12th-grade children. *Journal of Consulting and Clinical Psychology, 66,* 579-585.

Earleywine, M. (1994). Personality risk for alcoholism and alcohol expectancies. *Addictive Behaviors, 19,* 577-582.

Easdon, C.M., and Vogel-Sprott, M. (2000). Alcohol and behavioral control: Impaired response inhibition and flexibility in social drinkers. *Experimental and Clinical Psychopharmacology, 8,* 387-394.

Eaton, D.K., Kann, L., Kinchen, S., Ross, J., Hawkins, J., Harris, W.A., Lowry, R., McManus, T., Chyen, D., Shanklin, S., Lim, C., Grunbaum, J.A., and Wechsler, H. (2006). Youth risk behavior surveillance – United States, 2005. *Morbidity Mortality Weekly Report Surveillance Summary, 55,* 1-108.

Elder, R.W., Shults, R.A., Swahn, M.H., Strife, B.J., and Ryan, G.W. (2004). Alcohol-related emergency department visits among people ages 13 to 25 years. *Journal of Studies on Alcohol, 65,* 297-300.

Ellickson, P.L., Collins, R.L., Hambarsoomians, K., and McCaffrey, D.F. (2005). Does alcohol advertising promote adolescent drinking? Results from a longitudinal assessment. *Addiction, 100,* 235-246.

Englund, M.M., Egeland, B., Oliva, E.M., and Collins, W.A. (2008). Childhood and adolescent predictors of heavy drinking and alcohol use disorders in early adulthood: a longitudinal developmental analysis. *Addiction, 103 Suppl 1,* 23-35.

Engs, R.C., Diebold, B.A., and Hansen, D.J. (1996). The drinking patterns and problems of a national sample of college students, 1994. *Journal of Alcohol and Drug Education, 41,* 13-33.

Ennet, S.T., Tobler, N.S., Ringwalt, C.L., and Flewelling, R.L. (1994). How effective is drug abuse resistance education? A meta-analysis of Project DARE outcome. *American Journal of Public Health, 84,* 1394-1401.

Evans, S.M., and Levin, F.R. (2004). Differential response to alcohol in light and moderate female social drinkers. *Behavioral Pharmacology, 15*, 167-181.

Everett, S.A., Lowry, R., Cohen, L.R., and Dellinger, A.M. (1999). Unsafe motor vehicle practices among substance-using college students. *Accident Analysis and Prevention, 31*, 667-673.

Ewing, J.A. (1984). Detecting alcoholism: The CAGE questionnaire. *Journal of the American Medical Association, 252*, 1905–1917.

Eysenck, H.J., and Eysenck, M.W. (1985). *Personality and Individual Differences: a Natural Science Approach.* New York, NY: Plenum.

Eysenck, S.B.G., Pearson, PR., Easting, G., and Allsop, J.F. (1985). Age norms for impulsiveness, venturesomeness and empathy in adults. *Personality and Individual Differences, 6*, 613-619.

Fabbri, A., Marchesini, G., Morselli-Labate, A.M., Rossi, F., Cicognani, A., Dente, M., Lervese, T., Ruggeri, S., Mengozzi, U., and Vandelli, A. (2002). Positive blood alcohol concentration and road accidents. A prospective study in an Italian emergency department. *Emergency Medicine Journal, 19*, 210-214.

Fadda, F., and Rossetti, Z. (1998). Chronic ethanol consumption: from neuroadaptation to neurodegeneration. *Progress in Neurobiology, 56*, 385-431.

Falk, J.L., and Feingold, D.A. (1987). Environmental and cultural factors in the behavioral action of drugs. In H.Y. Meltzer (Ed.), *The Third Generation of Progress* (pp. 1503-1510). New York, NY: Raven Press.

Fan, A.Z., Russell, M., Naimi, T., Li, Y., Liao, Y., Jiles, R., and Mokdad, A.H. (2008). Patterns of alcohol consumption and the metabolic syndrome. *Journal of Clinical Endocrinology and Metabolism, in press.*

Fell, J.C., Fisher, D.A., Voas, R.B., Blackman, K., and Tippetts, A.S. (2008). The relationship of underage drinking laws to reductions in drinking drivers in fatal crashes in the United States. *Accident Analysis and Prevention, 40*, 1430-1440.

Ferreira, S.E., de Mello, M.T., and Formigoni, M.L. (2004). Can energy drinks affect the effects of alcoholic beverages? A study with users. *Revista da Associacao Medica Brasileira, 50*, 48-51.

Ferreira, S.E., de Mello, M.T., Pompeia, S., and de Souza-Formigoni, M.L.O. (2006). Effects of energy drink ingestion on alcohol intoxication. *Alcoholism: Clinical and Experimental Research, 30*, 598-605.

Ferreira, S.E., de Mello, M.T., Rossi, M.V., and Souza-Formigon, M.L.O. (2004). Does an energy drink modify the effects of alcohol in a maximal effort test? *Alcoholism: Clinical and Experimental Research, 28*, 1408-1412.

Feldman, L., Harvey, B., Holowaty, P., and Shortt, L. (1999). Alcohol use beliefs and behaviors among high school students. *Journal of Adolescent Health, 24*, 48-58.

Fell, J.C., and Voas, R.B. (2006). The effectiveness of reducing illegal blood alcohol concentration (BAC) limits for driving: Evidence for lowering the limit to .05 BAC. *Journal of Safety Research, 37*, 233-243.

Ferreira, S.E., Hartmann Quadros, I.M., Trindade, A.A., Takahashi, S., Koyama, R.G., Souza-Formigoni, M.L.O. (2004). Can energy drinks reduce the depressor effect of ethanol? An experimental study in mice. *Physiology and Behavior, 82*, 841-847.

Field, C.A., and O'Keefe, G. (2004). Behavioral and psychological risk factors for traumatic injury. *The Journal of Emergency Medicine, 26*, 27-35.

Fillmore, M.T. (2001). Cognitive preoccupation with alcohol and binge drinking in college students: Alcohol-induced priming of the motivation to drink. *Psychology of Addictive Behaviors, 15*, 325-332.

Fillmore, M.T. (2003). Drug abuse as a problem of impaired control: Current approaches and findings. *Behavioral and Cognitive Neuroscience Reviews, 2*, 179-197.

Fillmore, M.T. (2003). Alcohol tolerance in humans is enhanced by prior caffeine antagonism of alcohol-induced impairment. *Experimental and Clinical Psychopharmacology, 11*, 9-17.

Fillmore, M.T., and Blackburn, J. (2002). Compensating for alcohol-induced impairment: Alcohol expectancies and behavioral disinhibition. *Journal of Studies on Alcohol, 63*, 237-246.

Fillmore, M.T., Marczinski, C.A., and Bowman, A.M. (2005). Acute tolerance to alcohol effects on inhibitory and activational mechanisms of behavioral control. *Journal of Studies on Alcohol, 66*, 663-672.

Fillmore, M.T., Roach, E.L., and Rice, J.T. (2002). Does caffeine counteract alcohol-induced impairment? The ironic effects of expectancy. *Journal of Studies on Alcohol, 63*, 745-754.

Fillmore, M.T., and Vogel-Sprott, M. (1996). Evidence that expectancies mediate behavioral impairment under alcohol. *Journal of Studies on Alcohol, 57*, 598-603.

Fillmore, M.T., and Vogel-Sprott, M. (1999). An alcohol model of impaired inhibitory control and its treatment in humans. *Experimental and Clinical Psychopharmacology, 7*, 49-55.

Fillmore, M.T., and Vogel-Sprott, M. (2000). Response inhibition under alcohol: Effects of cognitive and motivational conflict. *Journal of Studies on Alcohol, 61*, 239-246.

Finn, P.R., Kessler, D.N., and Hussong, A.M. (1994). Risk for alcoholism and classical conditioning to signals for punishment: Evidence for a weak behavioral inhibition system? *Journal of Abnormal Psychology, 103*, 293-301.

Fisher, L.B., Miles, I.W., Austin, S.B., Camargo, C.A., and Colditz, G.A. (2007). Predictors of initiation of alcohol use among U.S. adolescents: findings from a prospective cohort study. *Archives of Pediatric and Adolescent Medicine, 161*, 959-966.

Fleming, K., Thorson, E., and Atkin, C.K. (2004). Alcohol advertising exposure and perceptions: links with alcohol expectancies and intentions to drink or drinking in underaged youth and young adults. *Journal of Health Communication, 9*, 3-29.

Flowers, N.T., Naimi, T.S., Brewer, R.D., Elder, R.W., Shults, R.A., and Jiles, R. (2008). Patterns of alcohol consumption and alcohol-impaired driving in the United States. *Alcoholism: Clinical and Experimental Research, 32*, 639-644.

Forchheimer, M., Cunningham, R.M., Gater, D.R. Jr., and Maio, R.F. (2005). The relationship of blood alcohol concentration to impairment severity in spinal cord injury. *Journal of Spinal Cord Medicine, 28*, 303-307.

Fournier, M.E., and Levy, S. (2006). Recent trends in adolescent substance use, primary care screening, and updates in treatment options. *Current Opinions in Pediatrics, 18*, 352-358.

Fowles, D.C. (1987). Application of a behavioral theory of motivation to the concepts of anxiety and impulsivity. *Journal of Research in Personality, 21*, 417-435.

Fox News (November 5, 2007). Alcohol, caffeinated energy drinks dangerous mix, study says. Retrieved September 5, 2008 from http://www.foxnews.com/0,3566,308076,00.html.

Frank, M.J., and Claus, E.D. (2006). Anatomy of a decision: Striato-orbitofrontal interactions in reinforcement learning, decision making, and reversal. *Psychological Bulletin, 113*, 300-326.

Franken, I.H., van Strien, J.W., Nijs, I., and Muris, P. (2008). Impulsivity is associated with behavioral decision-making deficits. *Psychiatry Research, 158*, 155-163.

Frazier, T.W., Youngstrom, E.A., Glutting, J.J., and Watkins, M.W. (2007). ADHD and achievement: Meta-analysis of the child, adolescent, and adult literatures and a concomitant study with college students. *Journal of Learning Disabilities, 40*, 49-65.

Fromme, K., and Samson, H.H. (1983). A survey analysis of first intoxication experiences. *Journal of Studies on Alcohol, 44*, 905-910.

Frunkel, E.N., Kanner, J., German, J.B., Parks, E., and Kinsella, J.E. (1993). Inhibition of oxidation of human low-density lipoprotein by phenolic substances in red wine. *Lancet, 341*, 454-457.

Gauthier, R. (2008). Miami student pleads to lesser charges in drinking death. *Middletown Journal, January 25, 2008.*

Gaziano, J.M., Buring, J.E., Breslow, J.L., Goldhaber, S.Z., Rosner, B., VanDenburgh, M., Willett, W., and Hennekens, C.H. (1993). Moderate alcohol intake, increased levels of high-density lipoprotein and its subfractions, and decreased risk of myocardial infarction. *New England Journal of Medicine, 329*, 1829-1834.

Gentilello, L.M., Rivara, F.P., Donovan, D.M., Jurkovich, G.J., Daranciang, E., Dunn, C.W., Villaveces, A., Copass, M., and Ries, R.R. (1999). Alcohol interventions in a trauma center as a means of reducing the risk of injury recurrence. *Annals of Surgery, 230*, 473-80.

Gentilello, L.M., Villaveces, A., Ries, R.R., Nason, K.S., Daranciang, E., Donovan, D.M., Copass, M., Jurkovich, G.J., and Rivara, F.P. (1999). Detection of acute alcohol intoxication and chronic alcohol dependence by trauma center staff. *Journal of Trauma, 47*, 1131-1135.

Giancola, P.R., Zeichner, A., Yarnell, J.E., and Dickson, K.E. (1996). Relation between executive cognitive functioning and the adverse consequences of alcohol use in social drinkers. *Alcoholism: Clinical and Experimental Research, 20*, 1094-1098.

Giedd, J.N., Snell, J.W., Lange, N., Rajapakse, J.C., Casey, C.J., Kozuch, P.L., Vaituzis, A.C., Vauss, Y.C., Hamburger, S.D., Kaysen, D., and Rapoport, J.L. (1996). Quantitative magnetic resonance imaging of human brain development: ages 4-18. *Cerebral Cortex, 6*, 551-560.

Gill, J.S. (2002). Reported levels of alcohol consumption and binge drinking within the UK undergraduate student population over the last 25 years. *Alcohol and Alcoholism, 37*, 109-120.

Gill, J.S., and Donaghy, M. (2004). Variation in the alcohol content of a 'drink' of wine and spirit poured by a sample of the Scottish population. *Health Education Research, 19*, 485-491.

Gillers, H., Jarosz, F., and McFeely, D. (2008). Student death may be 2nd linked to drinking: Some question whether Wabach College needs more alcohol education. *IndyStar.com*, retrieved October 11, 2008, from http://www.indystar.com.

Gliksman, L., Adlaf, E.M., Demers, A., and Newton-Taylor, B. (2003). Heavy drinking on Canadian campuses. *Canadian Journal of Public Health, 94*, 17-21.

Gmel, G., Bissery, A., Gammeter, R., Givel, J.C., Calmes, J.M., Yersin, B., and Daeppen, J.B. (2006). Alcohol-attributable injuries in admissions to a swiss emergency room – an analysis of the link between volume of drinking, drinking patterns, and preattendance drinking. *Alcoholism: Clinical and Experimental Research, 30*, 501-509.

Gmel, G., Givel, J.C., Yersin, B., and Daeppen, J.B. (2007). Injury and repeated injury – what is the link with acute consumption, binge drinking and chronic heavy alcohol use? *Swiss Medical Weekly, 137*, 642-648.

Goldman, M.S., Brown, S.A., Christiansen, B.A., and Smith, G.T. (1991). Alcoholism etiology and memory: Broadening the scope of alcohol expectancy research. *Psychological Bulletin, 110*, 137-146.

Goudriaan, A.E., Grekin, E.R., and Sher, K.J. (2007). Decision making and binge drinking: A longitudinal study. *Alcoholism: Clinical and Experimental Research, 31*, 928-938.

Graham, K., Leonard, K.E., Room, R., Wild, C., Pihl, R.P., Boiss, C., and Single, E. (1998). Current directions in research on understanding and preventing intoxicated aggression. *Addiction, 93*, 659-676.

Graham, D.M., Maio, R.F., Blow, F.C., and Hill, E.M. (2000). Emergency physician attitudes concerning intervention for alcohol abuse/dependence delivered in the emergency department: a brief report. *Journal of Addictive Diseases, 19*, 45-53.

Gray, J.A. (1976). The behavioral inhibition system: A possible substrate for anxiety. In M.P. Feldman and A. Broadhurst (Eds.), *Theoretical and Experimental Bases of the Behavior Therapies* (pp. 3-41). London: Wiley.

Gray. J.A. (1977). Drug effects on fear and frustration. Possible limbic site of action of minor tranquilizers. In L.L. Iverson, S.D. Iverson, and S.H. Snyder (Eds.), *Handbook of Psychopharmacology* (Vol. 8, pp/433-529). New York, Plenum.

Greenfield, T.K., and Room, R. (1997). Situational norms for drinking and drunkenness: trends in the US population, 1979-1990. *Addiction, 92*, 33-47.

Griffin, K.W., Botvin, G.J., Epstein, J.A., Doyle, M.M., and Diaz, T. (2000). Psychosocial and behavioral factors in early adolescence as predictors of heavy drinking among high school seniors. *Journal of Studies on Alcohol, 61*, 603-606.

Gronbaek, M., Deis, A., Sorensen, T.I.A., Becker, U., Schnohr, P., and Jensen, G. (1995). Mortality associated with moderate intake of wine, beer, or spirits. *British Medical Journal, 310*, 1165-1169.

Grube, J.W., and Wallack, L. (1994). Television beer advertising and drinking knowledge, beliefs, and intentions among schoolchildren. *American Journal of Public Health, 84*, 254-259.

Grucza, R.A., Norberg, K., Bucholz, K.K., and Bierut, L.J. (2008). Correspondence between secular changes in alcohol dependence and age of drinking onset among women in the United States. *Alcoholism: Clinical and Experimental Research, 32*, 1493-1501.

Guo, J., Chung, I.J., Hill, K.G., Hawkins, J.D., Catalano, R.F., and Abbott, R.D. (2002). Developmental relationships between adolescent substance use and risky sexual behavior in young adulthood. *Journal of Adolescent Health, 31*, 354-362.

Hanes, J.C., Farrell, M., Singleton, N., Meltzer, H., Araya, R., Lewis, G., and Wiles, N.J. (2008). Alcohol consumption as a risk factor for non-recovery from common mental disorder: results from the longitudinal follow-up of the National Psychiatric Morbidity Survey. *Psychological Medicine, 38*, 451-455.

Hariri, A.R., Bookheimer, S.Y., and Mazziotta, J.C. (2000). Modulating emotional responses: effects of a neocortical network on the limbic system. *Neuroreport, 11*, 43-48.

Harrison, E.L.R., and Fillmore, M.T. (2005a). Are bad drivers more impaired by alcohol? Sober driving precision predicts impairment from alcohol in a simulated driving task. *Accident Analysis and Prevention, 37*, 882-889.

Harrison, E.L.R., and Fillmore, M.T. (2005b). Social drinkers underestimate the additive impairing effects of alcohol and visual degradation on behavioral functioning. *Psychopharmacology, 177*, 459-464.

Harrison, E.L.R., Marczinski, C.A., and Fillmore, M.T. (2007). Driver training conditions affect sensitivity to the impairing effects of alcohol on a simulated driving test. *Experimental and Clinical Psychopharmacology, 15*, 288-298.

Harrison, E.L.R., and McKee, S.A. (2008). Young adult non-daily smokers: patterns of alcohol and cigarette use. *Addictive Behavior, 33*, 668-674.

Hartley, D.E., Elsabagh, S., and File, S.E. (2004). Binge drinking and sex: effects of mood and cognitive function in healthy young volunteers. *Pharmacology, Biochemistry and Behavior, 78*, 611-619.

Heath, D.B. (Ed.). (1995). *International handbook on alcohol and culture*. Westport, CT: Greenwood Press.

Heath, D.B. (1998). Cultural variations in drinking patterns. In M.Grant and J. Litvak (Eds.), *Drinking patterns and their consequences* (pp. 102-125). Philadelphia: Taylor and Francis.

Heath, D.B. (2000). *Drinking occasions: Comparative perspectives on alcohol and culture*. Philadelphia, PA: Brunner/Mazel.

Helmkamp, J.C., Hungerford, D.W., Williams, J.M., Manley, W.G., Furbee, P.M., Horn, K.A., and Pollock, D.A. (2003). Screening and brief intervention for alcohol problems among college students treated in a university hospital emergency department. *Journal of American College Health, 52*, 7-16.

Henriksen, L., Feighery, E.C., Schleicher, N.C., Fortmann, S.P. (2008). Receptivity to alcohol marketing predicts initiation of alcohol use. *Journal of Adolescent Health, 42*, 28-35.

Henry, F.P., Purcell, E.M., and Eadie, P.A. (2007). The human bite injury: a clinical audit and discussion regarding the management of this alchol fuelled phenomenon. *Emergency Medicine Journal, 24*, 455-458.

Hildebrandt, H., Brokate, B., Eling, P., and Lanz, M. (2004). Response shifting and inhibition, but not working memory, are impaired after long-term heavy alcohol consumption. *Neuropsychology, 18*, 203-211.

Hiltunen, A.J. (1997). Acute alcohol tolerance in social drinkers: changes in subjective effects dependent on the alcohol dose and prior alcohol experience. *Alcohol, 14*, 373-378.

Hiltunen, A.J., and Jarbe, T.U.C. (1990). Acute tolerance to ethanol using drug discriminiation and open-field procedures in rats. *Psychopharmacology, 102*, 207-212.

Hingson, R.W., Heeren, T., Jamanka, A., and Howland, J. (2000). Age of drinking onset and unintentional injury involvement after drinking. *Journal of the American Medical Association, 284*, 1527-1533.

Hingson, R., Heeren, T., and Winter, M. (1994). Lower legal blood alcohol limits for young drivers. *Public Health Reports, 109*, 738-744.

Hingson, R., Heeren, T., Winter, M. and Wechsler, H. (2005). Magnitude of alcohol-related mortality and morbidity among U.S. college students ages 18-24: Changes from 1998 to 2001. *Annual Review of Public Health, 26*, 259-279.

Hingson, R.W., Heeren, T., Zakocs, R.C., Kopstein, A., and Wechsler, H. (2002). Magnitude of alcohol-related mortality and morbidity among U.S. college students ages 18-24. *Journal of Studies on Alcohol, 63*, 136-144.

Hingson, R.W., McGovern, T., Howland, J., Heeren, T., Winter, M., and Zakocs, R. (1996). Reducing alcohol-impaired driving in Massachusetts: the Saving Lives Program. *American Journal of Public Health, 86*, 791-797.

Hinson, R.E. (1985). Individual differences in tolerance and relapse: A Pavlovian conditioning perspective. In M. Galizio and S.A. Maisto (Eds.), *Determinants of Substance Abuse: Biological, Psychological, and Environmental Factors* (pp. 101-124). New York, NY: Plenum Press.

Holdstock, L., King, A.C., de Wit, H. (2000). Subjective and objective responses to ethanol in moderate/heavy and light social drinkers. *Alcoholism: Clinical and Experimental Research, 24*, 789-794.

Horn, K., Leontieva, L., Williams, J.M., Furbee, P.M., Helmkamp, J.C., and Manley, W.G. 3rd (2002). Alcohol problems among young adult emergency department patients: making predictions using routine sociodemographic information. *Journal of Critical Care, 17*, 212-220.

Horne, J.A., and Reyner, L.A. (2001). Beneficial effects of an "energy drink" given to sleepy drivers. *Amino Acids, 20*, 83-89.

Hover, A.R., Hover, B.A., and Young, J.C. (2000). Measuring the effectiveness of a community sponsored DWI intervention for teens. *American Journal of Health Studies, 16*, 171-177.

Hull, J.G., and Bond, C.F. (1986). Social and behavioral consequences of alcohol consumption: A meta analysis. *Psychological Bulletin, 99*, 347-360.

Hulse, G.K., Robertson, S.I., and Tait, R.J. (2001). Adolescent emergency department presentations with alcohol- or other drug-related problems in Perth, Western Australia. *Addiction, 96*, 1059-1067.

Humphrey, G., Casswell, S., and Han, D.Y. (2003). Alcohol and injury among attendees at a New Zealand emergency department. *The New Zealand Medical Journal, 116*, U298.

Hunt, W.A. (1993). Are binge drinkers more at risk of developing brain damage? *Alcohol, 10*, 559-561.

International Center for Alcohol Policies (ICAP) (retrieved September 3, 2008). Blood Alcohol Concentration Limits Worldwide. Table 1. Standard BAC Limits. http://www.icap.org.

ICAP (retrieved September 25, 2008). Minimum Age Limits Worldwide. Table. http://www.icap.org.

Inzlicht, M., and Gutsell, J.N. (2007). Running on empty: neural signals for self-control failure. *Psychological Science, 18*, 933-937.

Jacobs, M.R., and Fehr, K. (1987). *Drugs and drug abuse: A reference text (2nd ed.).* Toronto, ON: Addiction Research Foundation.

Jamison, J., and Myers, L.B. (2008). Peer-group and price influence students drinking along with planned behaviour. *Alcohol and Alcoholism, 43*, 492-497.

Jellinek, E.M. (1952). Phases of alcohol addiction. *Quarterly Journal of Studies of Alcohol, 13*, 673-684.

Jentsch, J.D., and Taylor, J.R. (1999). Impulsivity resulting from fronto-striatal dysfunction in drug abuse: Implication for the control of behavior by reward-related stimuli. *Psychopharmacology, 146*, 373-390.

Jernigan, T.L., Trauner, D.A., Hesselink, J.R., and Tallal, P.A. (1991). Maturation of human cerebrum observed in vivo during adolescence. *Brain, 114*, 2037-2049.

Johnston, J.J., and McGovern, S.J. (2004). Alcohol related falls: an interesting pattern of injuries. *Emergency Medicine Journal, 21*, 185-188.

Johnston, L.D., O'Malley, P.M., and Bachman, J.G. (1991). *Drug Use Among American High School Seniors, College Students, and Young Adults, 1975-1990, Vol. 2.* Washington, DC: Government Printing Office. US Department of Health and Human Services publication ADM 91-1835.

Johnston, L.D., O'Malley, P.M., Bachman, J.G., and Schulenberg, J.E. (2007). *Monitoring the Future National Results on Adolescent Drug Use: Overview of Key Findings, 2006.* (NIH Publication No. 07-6202). Bethesda, MD: National Institute on Drug Abuse.

Jones, M.C. (1968). Personality correlates and antecedents of drinking patterns in adult males. *Journal of Consulting and Clinical Psychology, 32*, 2-12.

Journal of Studies on Alcohol and Drugs (retrieved August 15, 2008). Guidance for authors on the policy of the Journal of Studies on Alcohol and Drugs regarding the appropriate use of the term "binge". http://www.jsad.com/jsad/static/binge.html.

Kalant, H., LeBlanc, A.E., and Gibbins, R.J. (1971). Tolerance to, and dependence on, some nonopiate psychotropic drugs. *Pharmacological Reviews, 23*, 135-191.

Kalat, J.W. (2008). *Introduction to Psychology* (8th ed.). Belmont, CA: Thomson Higher Education.

Kasenda, M., Calzavara, L.M., Johnson, I., and LeBlanc, M. (1997). Correlates of condom use in the young adult population in Ontario. *Canadian Journal of Public Health, 88*, 280-285.

Kelly, T.M., Donovan, J.E., Chung, T., Cook, R.L., and Delbridge, T.R. (2004). Alcohol use disorders among emergency department-treated older adolescents: a new brief screen (RUFT-Cut) using AUDIT, CAGE, CRAFFT, and RAPS-QF. *Alcoholism: Clinical and Experimental Research, 28*, 746-753.

Kelly, T.M., Donovan, J.E., Cornelius, J.R., and Delbridge, T.R. (2004). Predictors of problem drinking among older adolescent emergency department patients. *Journal of Emergency Medicine, 27*, 209-218.

Kendler, K.S., Heath, A.C., Neale, M.C., Kessler, R.C., and Eaves, L.J. (1992). A population-based study of alcoholism in women. *Journal of the American Medical Association, 268*, 1877-1882.

Kerr, W.C., and Greenfield, T.K. (2003). The average ethanol content of beer in the U.S. and individual states: estimates for use in aggregate consumption statistics. *Journal of Studies on Alcohol, 64*, 70-74.

Kessler, R.C., Adler, L., Ames, M., Barkley, R.A., Birnbaum, H., and Greenberg, P. (2005). The prevalence and effects of adult attention deficiency/hyperactivity disorder on work performance in a nationally representative sample of workers. *Journal of Occupational and Educational Medicine, 47*, 565-572.

King, A.C., and Byars, J.A. (2004). Alcohol-induced performance impairment in heavy episodic and light social drinkers. *Journal of Studies on Alcohol, 65,* 27-36.

King, A.C., Houle, T., de Wit, H., Holdstock, L., and Schuster, A. (2002). Biphasic alcohol response differs in heavy versus light drinkers. *Alcoholism: Clinical and Experimental Research, 26,* 827-835.

Kinney, J. (2009). *Loosening the Grip: A Handbook of Alcohol Information.* (9th ed.). New York, NY: McGraw Hill.

Kitano, H.H.L. (1989). Alcohol and the Asian American. In T.D. Watts and R. Wright, Jr. (Eds.), *Alcoholism in Minority Populations* (pp. 143-158). Springfield, IL: Charles C. Thomas.

Kivlahan, D.R., Marlatt, G.A., Fromme, K., Coppel, D.B., and Williams, E. (1990). Secondary prevention with college drinkers: Evaluation of an alcohol skills training program. *Journal of Consulting and Clinical Psychology, 58,* 805-810.

Knight, J.R., Wechsler, H., Kuo, M., Seibring, M., Weitzman, E.R., and Schuckit, M. (2002). Alcohol abuse and dependence among U.S. college students. *Journal of Studies on Alcohol, 63,* 263-270.

Krain, A.L., and Castellanos, F.X. (2006). Brain development and ADHD. *Clinical Psychology Review, 26,* 433-444.

Kraus, C.L., Salazar, N.C., Mitchell, J.R., Florin, W.D., Guenther, B., Brady, D., Swartzwelder, S.H., and White, A.M. (2005). Inconsistencies between actual and estimated blood alcohol concentrations in a field study of college students: Do students really know much they drink? *Alcoholism: Clinical and Experimental Research, 29,* 1672-1676.

Kunin, D., Gaskin, S., Rogan, F., Smith, B.R., and Amit, Z. (2000). Caffeine promotes ethanol drinking in rates. Examination using a limited-access free choice paradigm. *Alcoholism, 21,* 271-277.

Kuo, M., Adlaf, E.M., Lee, H., Gliksman, L., Demers, A., and Wechsler, H. (2002). More Canadian students drink but American students drink more: comparing college alcohol use in two countries. *Addiction, 97,* 1583-1592.

Lack, A.K., Diaz, M.R., Chappell, A., DeBois, D.W., and McCool, B.A. (2007). Chronic ethanol and withdrawal differentially modulate pre- and postsynaptic function at glutamatergic synapses in rat basolateral amygdala. *Journal of Neurophysiology, 98,* 3185-3196.

Lange, J.E., Lauer, E.M., and Voas, R.B. (1999). A survey of the San Diego-Tijuana cross-border binging. Methods and analysis. *Evaluation Review, 23,* 378-398.

Lange, J.E., and Voas, R.B. (2000). Youth escaping limits on drinking: binging in Mexico. *Addiction, 95,* 521-528.

Lange, J.E., Voas, R.B., and Johnson, M.B. (2002). South of the border: a legal haven for underage drinking. *Addiction, 97,* 1195-1203.

Las Vegas Sun Editorial (2008). A cruel hoax: Twist on popular school program leads to unwarranted trauma. *Las Vegas Sun, June 18, 2008.*

Leigh, B.C., and Stacy, A.W. (2004). Alcohol expectancies and drinking in different age groups. *Addiction, 99,* 215-227.

Leonard, T.K., Watson, R.R., and Mohs, M.E. (1987). The effects of caffeine in various body systems: A review. *Journal of the American Dietetic Association, 87,* 1048-1053.

Leontieva, L., Horn, K., Haque, A., Helmkamp, J., Ehrlich, P., and Williams, J. (2005). Readiness to change problematic drinking assessed in the emergency department as a predictor of change. *Journal of Critical Care, 20*, 251-256.

Li, Y.M. (2007). Deliberate self-harm and relationship to alcohol use at an emergency department in eastern Taiwan. *The Kaohsiung Journal of Medical Sciences, 23*, 247-253.

Li, G., Baker, S.P., Smialek, J.E., and Soderstrom, C.A. (2001). Use of alcohol as a risk factor for bicycling injury. *Journal of the American Medical Association, 285*, 893-896.

Li, G., Chanmugam, A., Rothman, R., DiScala, C., Paidas, C.N., and Kelen, G.D. (1999). Alcohol and other psychoactive drugs in trauma patients aged 10-14 years. *Injury Prevention, 5*, 94-97.

Little, P.J., Kuhn, C.M., Wilson, W.A., and Swartzwelder, H.S. (1996). Differential effects of ethanol in adolescent and adult rats. *Alcoholism: Clinical and Experimental Research, 20*, 1346-1351.

Logan, G.D. (1994). On the ability to inhibit thought and action: A user's guide to the stop-signal paradigm. In D. Dagenbach and T.H. Carr (Eds.), *Inhibitory Processes in Attention, Memory, and Language* (pp. 189-239). San Diego, CA: Academic Press.

Logan, G.D., and Cowan, W.B. (1984). On the ability to inhibit thought and action: A theory of an act of control. *Psychological Review, 91*, 295-327.

Logan, G.D., Cowan, W.B., and Davis, K.A. (1984). On the ability to inhibit simple and choice reaction time responses: A model and a method. *Journal of Experimental Psychology: Human Perception and Performance, 10*, 276-291.

Loiselle, J.M., Baker, M.D., Templeton, J.M. Jr., Schwartz, G., and Drott, H. (1993). Substance abuse in adolescent trauma. *Annals of Emergency Medicine, 22*, 1530-1534.

London, J.A., and Battistella, F.D. (2007). Testing for substance use in trauma patients: Are we doing enough? *Archives of Surgery, 142*, 633-638.

Longabaugh, R., Woolard, R.E., Nirenberg, T.D., Minugh, A.P., Becker, B., Clifford, P.R., Carty, K., Sparadeo, F., and Gogineni, A. (2001). Evaluating the effects of a brief motivational intervention for injured drinkers in the emergency department. *Journal of Studies on Alcohol, 62*, 806-816.

Luczak, S.E., Wall, T.L., Shea, S.H., Byun, S.M., and Carr, L.G. (2001). Binge drinking in Chinese, Korean, and White college students: genetic and ethnic group differences. *Psychology of Addictive Behaviors, 15*, 306-309.

Ludwig, A.M., Wikler, A., and Stark, L.H. (1974). The first drink: Psychobiological aspects of craving. *Archives of General Psychiatry, 30*, 539-547.

Luke, L.C., Dewar, C., Bailey, M., McGreevy, D., Morris, H., and Burdett-Smith, P. (2002). A little nightclub medicine: the healthcare implications of clubbing. *Emergency Medicine Journal, 19*, 542-545.

Lynam, D.R., Milich, R., Zimmerman, R., Novak, S.P., Logan, T.K., Martin, C. et al. (1999). Project DARE: No effects at 10-year follow-up. *Journal of Consulting and Clinical Psychology, 67*, 590-593.

MacAndrew, C., and Egerton, R.B. (1969). *Drunken comportment.* Chicago, IL: Aldine.

Magid, V., Maclean, M.G., and Colder, C.R. (2007). Differentiating between sensation seeking and impulsivity through their mediated relations with alcohol use and problems. *Addictive Behavior, 32*, 2046-2061.

Maio, R.F., Shope, J.T., Blow, F.C., Gregor, M.A., Zakrajsek, J.S., Weber, J.E., and Nypaver, M.M. (2005). A randomized controlled trial of an emergency department-based interactive computer program to prevent alcohol misuse among injured adolescents. *Annals of Emergency Medicine, 45*, 430-432.

Maisto, S.A., Galizio, M., and Connors, G.J. (2008). *Drug Use and Abuse (5ᵗʰ ed.)*. Belmont, CA: Thomson Higher Education.

Malinauskas, B.M., Aeby, V.G., Overton, R.F., Carpenter-Aeby, T., and Barber-Heidal (2007). A survey of energy drink consumption patterns among college students. *Nutrition Journal, 6*, 35.

Mann, L.M., Chassin, L., and Sher, K.J. (1987). Alcohol expectancies and the risk for alcoholism. *Journal of Consulting and Clinical Psychology, 55*, 411-417.

Mannuzza, S., Klein, R.G., Bessler, A., Malloy, P., and Hynes, M.E. (1997). Educatoinal and occupational outcomes of hyperactive boys grown up. *Journal of the American Academy of Child Psychiatry, 36*, 1222-1227.

Marczinski, C.A., Bryant, R., and Fillmore, M.T. (2005). The relationship between cognitive preoccupation with alcohol and alcohol use in male and female college students. *Addiction Research and Theory, 13*, 383-394.

Marczinski, C.A., Combs, S.W., and Fillmore, M.T. (2007). Increased sensitivity to the disinhibiting effects of alcohol in binge drinkers. *Psychology of Addictive Behaviors, 21*, 346-354.

Marczinski, C.A., and Fillmore, M.T. (2003a). Dissociative antagonistic effects of caffeine on alcohol-induced impairment of behavioral control. *Experimental and Clinical Psychopharmacology, 11*, 228-236.

Marczinski, C.A., and Fillmore, M.T. (2003b). Pre-response cues reduce the impairing effects of alcohol on the execution and suppression of responses. *Experimental and Clinical Psychopharmacology, 11*, 110-117.

Marczinski, C.A., and Fillmore, M.T. (2005a). Alcohol increases reliance on cues that signal acts of control. *Experimental and Clinical Psychopharmacology, 13*, 15-24.

Marczinski, C.A., and Fillmore, M.T. (2005b). Compensating for alcohol-induced impairment of control: Effects on inhibition and activation of behavior. *Psychopharmacology, 181*, 337-346.

Marczinski, C.A., and Fillmore, M.T. (2006). Clubgoers and their trendy cocktails: Implications for mixing caffeine into alcohol on information processing and subjective reports of intoxication. *Experimental and Clinical Psychopharmacology, 14*, 450-458.

Marczinski, C.A., and Fillmore, M.T. (in press). Acute alcohol tolerance on subjective intoxication and simulated driving performance in binge drinkers. *Psychology of Addictive Behaviors*.

Marczinski, C.A., Harrison, E.L.R., and Fillmore, M.T. (2008). Effects of alcohol on simulated driving and perceived driving impairment in binge drinkers. *Alcoholism: Clinical and Experimental Research, 32*, 1329-1337.

Margulies, R.Z., Kessler, R.C., and Kandel, D.B. (1977). A longitudinal study of onset of drinking among high-school students. *Journal of Studies on Alcohol, 38*, 897-912.

Markwiese, B.J., Acheson, S.K., Levin, E.D., Wilson, W.A., and Swartzwelder, H.S. (1998). Differential effects of ethanol on memory in adolescent and adult rats. *Alcoholism: Clinical and Experimental Research, 22*, 416-421.

Marlatt, G.A., Demming, B., and Reid, J.B. (1973). Loss of control drinking in alcoholics: An experimental analogue. *Journal of Abnormal Psychology, 81*, 233-241.

Marlatt, G.A., and Gordon, J.R. (1980). Determinants of relapse: Implications for the maintenance of behavior change. In P.O. Davidson and S.M. Davidson (Eds.), *Behavioral Medicine: Changing Health Lifestyles* (pp. 410-452). New York: Brunner/Mazel.

Marmot, M.G. (2001). Alcohol and coronary heart disease. *International Journal of Epidemiology, 30*, 724-729.

Martin, M. (1998). The use of alcohol among NCAA Division I female college basketball, softball, and volleyball athletes. *Journal of Athletic Training, 33*, 163-167.

Martin, C.S., Earleywine, M., Musty, R.E., Perrine, M.W., and Swift, R.M. (1993). Development and validation of the Biphasic Alcohol Effects Scale. *Alcoholism: Clinical and Experimental Research, 17*, 140-146.

Maugh, T.H. (2006). Was it alcohol or anti-Semitism talking? Doctors disagree. *Los Angeles Times.*

McCarthy, D.M., Miller, T.L., Smith, G.T., and Smith, J.A. (2001). Disinhibition and expectancy in risk for alcohol use: comparing black and white college samples. *Journal of Studies on Alcohol, 62*, 313-321.

McCusker, R.R., Goldberger, B.A., and Cone, E.J. (2006). Caffeine content of energy drinks, carbonated soda, and other beverages. *Journal of Analytical Toxicology, 30*, 112-114.

McDonald, A.J. 3rd, Wang, N., and Camargo, C.A. Jr. (2004). U.S. emergency department visits for alcohol-related diseases and injuries between 1992 and 2000. *Archives of Internal Medicine, 164*, 531-537.

McKinney, A., and Coyle, K. (2004). Next day effects of a normal night's drinking on memory and psychomotor performance. *Alcohol and Alcoholism, 39*, 509-513.

Mehrabian, A., and Russell, J.A. (1978). A questionnaire measure of habitual alcohol use. *Psychological Reports, 43*, 803-806.

Mello, M.J., Nirenberg, T.D., Longabaugh, R., Woolard, R., Minugh, A., Becker, B., Baird, J., and Stein, L. (2005). Emergency department brief motivational interventions for alcohol with motor vehicle crash patients. *Annals of Emergency Medicine, 45*, 620-625.

Meropol, S.B., Moscati, R.M., Lillis, K.A., Ballow, S., and Janicke, D.M. (1995). Alcohol-related injuries among adolescents in the emergency department. *Annals of Emergency Medicine, 26*, 221-223.

Michaelis, V. (2004). Olympic star Michael Phelps charged with drunken driving. *USA Today,* retrieved October 11, 2008, from http://www.usatoday.com/sports/olympics/summer/2004-11-08-phelps-arrest_x.html.

Miller, K.E. (2008). Wired: energy drinks, jock identity, masculine norms, and risk taking. *Journal of American College Health, 56*, 481-489.

Miller, J., Schaffer, R., and Hackley, S.A. (1991). Effects of preliminary information in a go versus no-go task. *Acta Psychologica, 76*, 241-292.

Miller, P.M., Smith, G.T., and Goldman, M.S. (1990). Emergence of alcohol expectancies in childhood: A possible critical period. *Journal of Studies on Alcohol, 51*, 343-349.

Mintel (2008). *Energy Drink Explosion Hits Food.* Press release retrieved September 8, 2008 from http://www.mintel.com/press_releases/391034.htm.

Mitchell, J.M., Fields, H.L., D'Esposito, M., and Boettiger, C.A. (2005). Impulsive responding in alcoholics. *Alcoholism: Clinical and Experimental Research, 29*, 2158-2169.

Monti, P.M., Barnett, N.P., Colby, S.M., Gwaltney, C.J., Spirito, A., Rohsenow, D.J., and Woolard, R. (2007). Motivational interviewing versus feedback only in emergency care for young adult problem drinking. *Addiction, 102*, 1234-1243.

Monti, P.M., Colby, S.M., Barnett, N.P., Spirito, A., Rohsenow, D.J., Myers, M., Woolard, R., and Lewander, W. (1999). Brief intervention for harm reduction with alcohol-positive older adolescents in a hospital emergency department. *Journal of Consulting and Clinical Psychology, 67*, 989-94.

Mooney, D.K., Fromme, K., Kivlahan, D.R., and Marlatt, G.A. (1987). Correlates of alcohol consumption: Sex, age, and expectancies relate differentially to quantity and frequency. *Addictive Behaviors, 12*, 235-240.

Moore, A.A., Giuli, L., Gould, R., Hu, P., Zhou, K., Reuben, D., Greendale, G., and Karlamangla, A. (2006). Alcohol use, comorbidity, and mortality. *Journal of the American Geriatrics Society, 54*, 757-762.

Moselhy, H.F., Georgiou, G., and Kahn, A. (2001). Frontal lobe change in alcoholism: a review of the literature. *Alcohol and Alcoholism, 36*, 357-368.

Moss, H.B. (2008). Special section: Alcohol and adolescent brain development. *Alcoholism: Clinical and Experimental Research, 32*, 427-429.

Mulvihill, L.E., Skilling, T.A., and Vogel-Sprott, M. (1997). Alcohol and the ability to inhibit behavior in men and women. *Journal of Studies on Alcohol, 58*, 600-605.

Miller, H.W., Naimi, T.S., Brewer, R.D., and Jones, S.E. (2007). Binge drinking and associated health risk behaviors among high school students. *Pediatrics, 119*, 76-85.

Naimi, T.S., Brewer, R.D., Mokdad, A., Denny, C., Serdula, M.K., and Marks, J.S. (2003). Binge drinking among US adults. *Journal of the American Medical Association, 289*, 70-75.

Naimi, T.S., Lipscomb, L.E., Brewer, R.D., and Gilbert, B.C. (2003). Binge drinking in the preconception period and the risk of unintended pregnancy: implications for women and their children. *Pediatrics, 111*, 1136-1141.

National Advisory Council on Alcohol Abuse and Alcoholism (2002). *A call to action: Changing the culture of drinking at U.S. colleges.* (NIH Publication No. 02-5010). Bethesda, MD: National Institute on Alcohol Abuse and Alcoholism.

National Highway Traffic Safety Adminstration (2002). Traffic safety facts 2002: Alcohol. Washington, DC: U.S. Department of Transportation, National Highway Traffic Safety Administration, DOT HS 809 606 (Available at www.nhtsa.dot.gov).

National Institute on Alcohol Abuse and Alcoholism (NIAAA) (2002). *A call to action: Changing the culture of drinking at U.S. colleges.* Washington, DC: U.S. Department of Health and Human Services.

National Institute of Alcohol Abuse and Alcoholism. (2004). *Helping patients with alcohol problems: a health practitioner's guide* (NIH publication no. 04-3769). Washington, DC: Government Printing Office.

NIAAA (2004, Winter). NIAAA council approves definition of binge drinking. *Newsletter, 3*, 3.

NIAAA (2008). National Minimum Drinking Age Act. *Alcohol Policy Information System.* Retrieved September 30, 2008, from www.alcoholpolicy.niaaa.nih.gov.

National Institute on Drug Abuse (1985). Consensus development panel: Drug concentrations and driving impairment. *Journal of the American Medical Association, 254*, 2618-2621.

Neighbors, C., Oster-Aaland, L., Bergstrom, R.L., and Lewis, M.A. (2006). Event- and context-specific normative misperceptions and high-risk drinking: 21st birthday celebrations and football tailgating. *Journal of Studies on Alcohol, 67*, 282-289.

Nelson, M.C., Lust, K., Story, M., and Ehlinger, E. (2008). Credit card debt, stress and key health risk behaviors among college students. *American Journal of Health Promotion, 22*, 400-407.

Nelson, T.F., Naimi, T.S., Brewer, R.D., and Wechsler, H. (2005). The state sets the rate: the relationship among state-specific college binge drinking, state binge drinking rates, and selected state alcohol control policies. *American Journal of Public Health, 95*, 441-446.

Nelson, T.F., and Wechsler, H. (2001). Alcohol and college athletes. *Medicine and Science in Sports and Exercise, 33*, 43-47.

Nelson, T.F., and Wechsler, H. (2003). School spirits: alcohol and collegiate sports fans. *Addictive Behaviors, 28*, 1-11.

Neumann, T., Neuner, B., Weiss-Gerlach, E., Tonnesen, H., Gentilello, L.M., Wernecke, K.D., Schmidt, K., Schroder, T., Wauer, H., Heinz, A., Mann, K., Muller, J.M., Haas, N., Kox, W.J., and Spies, C.D. (2006). The effect of computerized tailored brief advice on at-risk drinking in subcritically injured trauma patients. *Journal of Trauma, 61*, 805-814.

Niemela, S., Sourander, A., Poikolainen, K., Helenius, H., Sillanmaki, L., Parkkola, K., Piha, J., Kumpulainen, K. Almqvist, F., and Moilanen, I. (2006). Childhood predictors of drunkenness in late adolescence among males: a 10-year population-based follow-up study. *Addiction, 101*, 512-521.

Nilsen, P., Holmqvist, M., Nordqvist, C., and Bendtsen, P. (2007). Frequency of heavy episodic drinking among nonfatal injury patients attending an emergency room. *Accident Analysis and Prevention, 39*757-766.

Noble, E.P. (1983). Social drinking and cognitive function: a review. *Substance and Alcohol Actions/Misuse, 4*, 205-216.

Noel, X., Van der Linden, M., d'Acremont, M., Bechara, A., Dan, B., Hanak, C., and Verbanck, P. (2007). Alcohol cues increase cognitive impulsivity in individuals with alcoholism. *Psychopharmacology, 192*, 291-298.

Noel, X., Van der Linden, M., d'Acremont, M., Colmant, M., Hanak, C., Pelc, I., Verbanck, P., and Becgara, A. (2005). Cognitive biases toward alcohol-related words and executive deficits in polysubstance abusers with alcoholism. *Addiction, 100*, 1302-1309.

Nordqvist, C., Homqvist, M., Nilsen, P., Bendtsen, P., and Lindqvist, K. (2006). Usual drinking patterns and non-fatal injury among patients seeking emergency care. *Public Health, 120*, 1064-1073.

Obernier, J.A., Bouldin, T.W., and Crews, F.T. (2002). Binge ethanol exposure in adult rats causes necrotic cell death. *Alcoholism: Clinical and Experimental Research, 26*, 547-557.

O'Brien, K.S., Hunter, J., Kypri, K., and Ali, A. (2008). Gender equality in university sportspeople's drinking. *Drug and Alcohol Review, 29*, 1-7.

O'Brien, M.C., McCoy, T.P., Champion, H., Mitra, A., Robbins, A., Teuschlser, H., Wolfson, M., DuRant, R.H. (2006). Single question about drunkenness to detect college students at risk for injury. *Academic Emergency Medicine, 13*, 629-636.

O'Brien, M.C., McCoy, T.P., Rhodes, S.D., Magoner, A., and Wolfson, M. (2008). Caffeinated cocktails: Energy drink consumption, high-risk drinking, and alcohol-related consequences among college students. *Academic Emergency Medicine, 15*, 453-460.

O'Connell, H., and Lawlor, B.A. (2005). Recent alcohol intake and suicidality – a neuropsychological perspective. *Irish Journal of Medical Science, 174*, 51-54.

Ogden, E.J., and Moskowitz, H. (2004). Effects of alcohol and other drugs on driver performance. *Traffic Injury Prevention, 5*, 185-198.

Ogurzsoff, S., and Vogel-Sprott, M. (1976). Low blood alcohol discrimination and self-titration skills of social drinkers with widely varied drinking habits. *Canadian Journal of Behavioural Science, 8*, 232-242.

O'Keefe, J.H., Bybee, K.A., and Lavie, C.J. (2007). Alcohol and cardiovascular health: the razor-sharp double-edged sword. *Journal of the American College of Cardiology, 50*, 1009-1014.

O'Malley, P.M., and Johnston, L.D. (2002). Epidemiology of alcohol and other drug use among American college students. *Journal of Studies on Alcohol, Suppl. 14*, 23-39.

Oosterlaan, J., and Sergeant, J.A. (1996). Inhibition in ADHD, aggressive, and anxious children: A biologically based model of child psychopathology. *Journal of Abnormal Child Psychology, 24*, 19-37.

Oteri, A., Salvo, F., Caputi, A.P., and Calapai, G. (2007). Intake of energy drinks in association with alcoholic beverages in a cohort of students of the school of medicine of the University of Messina. *Alcoholism: Clinical and Experimental Research, 31*, 1677-1680.

Park, A., Sher, K.J., and Krull, J.L. (2008). Risky drinking in college changes as fraternity/sorority affiliation changes: a person-environment perspective. *Psychology of Addictive Behaviors, 22*, 219-229.

Parker-Pope, T. (May 27, 2008). Energy drinks linked to risky behavior among teenagers. *International Herald Tribune*. Retrieved online September 5, 2008, from http://www.iht.com.

Parrott, D.J., and Giancola, P.R. (2004). A further examination of the relation between trait anger and alcohol-related aggression: The role of anger control. *Alcoholism: Clinical and Experimental Research, 28*, 855-864.

Parrott, D.J., and Giancola, P.R. (2006). The effect of past-year heavy drinking on alcohol-related aggression. *Journal of Studies on Alcohol, 67*, 122-130.

Parry, C.D., Bhana, A., Myers, B., Pluddemann, A., Flisher, A.J., Peden, M.M., and Morojele, N.K. (2002). Alcohol use in South Africa: findings from the South African Community Epidemiology Network on Drug use (SACENDU) project. *Journal of Studies on Alcohol, 63*, 430-435.

Parsons, O.A., and Nixon, S.J. (1998). Cognitive functioning in sober social drinkers: a review of the research since 1986. *Journal of Studies on Alcohol, 59*, 180-190.

Pasch, K.E., Perry, C.L., Stigler, M.H., and Komro, K.A. (2008). Sixth grade students who use alcohol: Do we need primary prevention programs for "tweens"? *Health Education and Behavior, epub ahead of print (update reference)*.

Pascual, M., Blanco, A.M., Cauli, O., Minarro, J., and Guerri, C. (2007). Intermittent ethanol exposure induces inflammatory brain damage and causes long-term behavioural alternations in adolescent rats. *European Journal of Neuroscience, 25*, 541-550.

Patterson, C.M., and Newman, J.P. (1993). Reflectivity and learning from aversive events: Toward a psychological mechanism for the syndromes of disinhibition. *Psychological Review, 100*, 716-736.

Patton, J.H., Stanford, M.S., and Barratt, E.S. (1995). Factor structure of the Barratt Impulsiveness Scale. *Journal of Clinical Psychology, 51*, 768-774.

Paus, T., Zijdenbos, A., Worsley, K., Collins, D.L., Blumenthal, J., Giedd, J.N., Rapoport, J.L., and Evans, A.C. (1999). Structural maturation of neural pathways in children and adolescents: in vivo study. *Science, 283*, 1908-1911.

Perkins, H.W. (2002). Social norms and the prevention of alcohol misuse in collegiate contexts. *Journal of Studies on Alcohol, Suppl. 14*, 164-172.

Perkins, H.W. (2007). Misperceptions of peer drinking norms in Canada: another look at the "reign of error" and its consequences among college students. *Addictive Behavior, 32*, 2645-2656.

Perkins, H.W., and Craig, D.W. (2006). A successful social norms campaign to reduce alcohol misuse among college student-athletes. *Journal of Studies on Alcohol, 67*, 880-889.

Pernanen, K. (1976). Alcohol and crimes of violence. In B. Kissin and H. Begleiter (Eds.), *The Biology of Alcoholism: Vol. 4. Social Aspects of Alcoholism* (pp. 351-444). New York: Plenum Press.

Pernanen, K. (1993). Research approaches in the study of alcohol-related violence. *Alcohol Health and Research World, 17*, 101-107.

Petridou, E., Zavitsanor, X., Dessypris, N., Frangakis, C., Madyla, M., Doxiadis, S., and Trichopoulos, D. (1997). Adolescents in high-risk trajectory: clustering of risk behavior and the origins of socioeconomic health differentials. *Preventative Medicine, 26*, 215-219.

Petry, N.M. (2001). Delay discounting of money and alcohol in actively using alcoholics, currently abstinent alcoholics, and controls. *Psychopharmacology, 154*, 243-250.

Petry, N.M., Kirby, K.N., and Krantzler, H.R. (2002). Effects of gender and family history of alcohol dependence on a behavioral task of impulsivity in healthy subjects. *Journal of Studies on Alcohol, 63*, 83-90.

Pfefferbaum, A., Kim, K.O., Zipursky, R.B., Mathalon, D.H., Lane, B., Ha, C.N., Rosenbloom, M.J., and Sullivan, E.V. (1992). Brain gray and white matter volume loss accelerates with aging in chronic alcoholics: a quantitative MRI study. *Alcoholism: Clinical and Experimental Research, 16*, 1078-1089.

Pincock, S. (2003). Binge drinking on rise in UK and elsewhere. Government report shows increases in alcohol consumption, cirrhosis, and premature deaths. *Lancet, 362*, 1126-1127.

Piombo, M., and Piles, M. (1996). The relationship between college females' drinking and their sexual behaviors. *Women's Health Issues, 6*, 221-228.

Pirkle, E.C., and Richter, L. (2006). Personality, attitudinal and behavioral risk profiles of young female binge drinkers and smokers. *Journal of Adolescent Health, 38*, 44-54.

Portans, I., White, J.M., and Staiger, P.K. (1989). Acute tolerance to alcohol: changes in subjective effects among social drinkers. *Psychopharmacology, 97*, 365-369.

Porter, R.S. (2000). Alcohol and injury in adolescents. *Pediatric Emergency Care, 16*, 316-320.

Posner, M.I. (1980). Orienting of attention. *Quarterly Journal of Experimental Psychology, 32*, 3-25.

Posner, M.I., Synder, C.R., and Davidson, B.J. (1980). Attention and the detection of signals. *Journal of Experimental Psychology, 109*, 160-174.

Presley, C.A., Leichliter, M.A., and Meilman, P.W. (1998). *Alcohol and Drugs on American College Campuses: A Report to College Presidents: Third in a series, 1995, 1996, 1997.* Carbondale, IL: Core Institute, Southern Illinois University.

Presley, C.A., Meilman, P.W., and Cashin, J.R. (1996). *Alcohol and Drugs on American College Campuses: Use, Consequences, and Perceptions of the Campus Environment, Vol. IV: 1992-1994.* Carbondale, IL: Core Institute, Southern Illinois University.

Presley, C.A., Meilman, P.W., Cashin, J.R., and Lyerla, R. (1996). *Alcohol and Drugs on American College Campuses: Use, Consequences, and Perceptions of the Campus Environment, Vol. III: 1991-1993.* Carbondale, IL: Core Institute, Southern Illinois University.

Presley, C.A., Meilman, P.W., and Leichliter, J.S. (2002). College factors that influence drinking. *Journal of Studies on Alcohol Supplement, 14,* 82-90.

Presley, C.A., Meilman, P.W., and Lyerla, R. (1993). *Alcohol and Drugs on American College Campuses: Use, Consequence, and Perceptions of the Campus Environment, Vol. I: 1989-1991.* Carbondale, IL: Core Institute, Southern Illinois University.

Pridemore, W.A. (2004). Weekend effects on binge drinking and homicide: the social connection between alcohol and violence in Russia. *Addiction, 99,* 1034-1041.

Prime Minister's Strategy Unit (2004). *Alcohol Harm Reduction Strategy for England.* Department of Health, London, UK.

Puljula, J., Savola, O., Tuomivaara, V., Pribula, J., and Hillbom, M. (2007). Weekday distribution of head traumas in patients admitted to the emergency department of a city hospital: effects of age, gender and drinking pattern. *Alcohol and Alcoholism, 42,* 474-479.

Quay, H.C. (1997). Inhibition and attention deficit hyperactivity disorder. *Journal of Abnormal Child Psychology, 25,* 7-13.

Quinlan, K.P., Brewer, R.D., Siegel, P., Sleet, D.A., Mokdad, A.H., Shults, R.A., and Flowers, N. (2005). Alcohol-impaired driving among U.S. adults, 1993-2002. *American Journal of Preventative Medicine, 28,* 346-350.

Rather, B.C., Goldman, M.S., Roehrich, L., and Brannick, M. (1992). Empirical modeling of an alcohol expectancy memory network using multidimensional scaling. *Journal of Abnormal Psychology, 101,* 174-183.

Reback, C.J., Shoptaw, S., and Grella, C.E. (2008). Methamphetamine use trends among street-recruited gay and bisexual males, from 1999 to 2007. *Journal of Urban Health, ahead of print.*

Reed, D.N. Jr., Wolf, B., Barber, K.R., Kotlowski, R., Montanez, N., Saxe, A., Coffey, D.C., Pollard, M., Fitzgerald, H.E., and Richardson, J.D. (2005). The stages of change questionnaire as a predictor of trauma patients most likely to decrease alcohol use. *Journal of the American College of Surgeons, 200,* 179-185.

Reifman, A., and Watson, W.K. (2003). Binge drinking during the first semester of college: continuation and desistance from high school patterns. *Journal of American College Health, 52,* 73-81.

Reinfurt, D., Williams, A., Wells, J., and Rodgman, E. (1996). Characteristics of drivers not using seat belts in a high belt use state. *Journal of Safety Research, 27,* 209-215.

Reis, J., Harned, I., and Riley, W. (2004). Young adult's immediate reaction to a personal alcohol overdose. *Journal of Drug Education, 34,* 235-245.

Renaud, S., and deLorgerd, M. (1992). Wine, alcohol, platelets, and the French paradox for coronary heart disease. *The Lancet, 339*, 1523-1526.

Rhodes, K.V., Lauderdale, D.S., Stocking, C.B., Howes, D.S., Roizen, M.F., and Levinson, W. (2001). Better health while you wait: a controlled trial of a computer-based intervention for screening and health promotion in the emergency department. *Annals of Emergency Medicine, 37*, 284-291.

Richardson, A., and Budd, T. (2003). Young adults, alcohol, crime and disorder. *Criminal Behavior and Mental Health, 13*, 5-16.

Riesselmann, B., Rosenbaum, F., and Schneider, V. (1996). Alcohol and energy drink: Can combined consumption of both beverages modify automobile driving fitness? *Blutalkohol, 33*, 201-208.

Rimm, E.B. (2000). Moderate alcohol intake and lower risk of coronary heart disease: Meta-analysis of effects on lipids and homeostatic factors. *Journal of the American Medical Association, 319*, 1523-1528.

Ripley, TL., O'Shea, M., and Stephens, D.N. (2003). Repeated withdrawal from ethanol impairs acquisition but not expression of conditioned fear. *European Journal of Neuroscience, 18*, 441-448.

Rivara, F.P., Gurney, J.G., Ries, R.K., Seguin, D.A., Copass, M.K., and Jurkovich, G.J. (1992). A descriptive study of trauma, alcohol, and alcoholism in young adults. *Journal of Adolescent Health, 13*, 663-667.

Roberto, M., Madamba, S.G., Stouffer, D.G., Parsons, LH., and Siggins, G.R. (2004). Increased GABA release in the central amygdala of ethanol-dependent rats. *Journal of Neuroscience, 24*, 10159-10166.

Robinson, T.E., and Berridge, K.C. (1993). The neural basis of drug craving: An incentive-sensitization theory of addiction. *Brain Research Reviews, 18*, 247-291.

Robinson, T.E., and Berridge, K.C. (2003). Addiction. *Annual Review of Psychology, 54*, 25-53.

Rodriguez, G. (2008). Wabash mourning loss of freshman student. *WISH TV8 Indiapolis*, Retrieved October 11, 2008, from http://www.wishtv.com.

Room, R. (2001). Intoxication and bad behaviour: Understanding cultural differences in the link. *Social Science and Medicine, 53*, 189-198.

Room, R., and Makela, K. (2000). Typologies of the cultural position of drinking. *Journal of Studies on Alcohol, 61*, 475-483.

Rose, A.K., and Grunsell, L. (2008). The subjective, rather than the disinhibiting, effects of alcohol are related to binge drinking. *Alcoholism: Clinical and Experimental Research, 32*, 1096-1104.

Rosenberg, H. (1993). Prediction of controlled drinking by alcoholics and problem drinkers. *Psychological Bulletin, 113*, 129-139.

Rubio, G., Jimenez, M., Rodriguez-Jimenez, R., Martinez, I., Avila, C., Ferre, F., Jimenez-Arriero, M.A., Ponce, G., and Palomo, T. (2008). The role of behavioral impulsivity in the development of alcohol dependence: A 4-year follow-up study. *Alcoholism: Clinical and Experimental Research, in press*.

Russ, N.W., Harwood, M.K., and Geller, E.S. (1986). Estimating alcohol impairment in the field: implications for drunken driving. *Journal of Studies on Alcohol, 47*, 237-240.

Rutledge, P., Park, A., and Sher, K.J. (2008). 21st birthday drinking: Extremely extreme. *Journal of Consulting and Clinical Psychology, 76*, 511-516.

Ryb, G.E., Dischinger, P.C., Kufera, J.A., and Read, K.M. (2006). Risk perception and impulsivity: association with risky behaviors and substance use disorders. *Accident Analysis and Prevention, 38*, 567-573.

Saitz, R. (2005). Unhealthy alcohol use. *New England Journal of Medicine, 352*, 596-607.

Samson, H.H., and Harris, R.A. (1992). Neurobiology of alcohol abuse. *Trends in Pharmacological Science, 13*, 206-211.

Saunders, J.B., Aasland, O.G., Babor, T.F., de la Fuente, J.R., and Grant, M. (1993). Development of the alcohol use disorders identification test (AUDIT): WHO collaborative project on early detection of persons with harmful alcohol consumption - II. *Addiction, 88*, 791–803.

Schachar, R., Tannock, R., Marriott, M., and Logan, G. (1995). Deficient inhibitory control in attention deficit hyperactivity disorder. *Journal of Abnormal Child Psychology, 23*, 411-437.

Schermer, C.R., Qualls, C.R., Brown, C.L., and Apodaca, T.R. (2001). Intoxicated motor vehicle passengers: an overlooked at-risk population. *Archives of Surgery, 136*, 1244-1248.

Schuckit, M.A. (1987). Biology of risk of alcoholism. In H.Y. Meltzer (Ed.), *Psychopharmacology: The Third Generation of Progress* (pp. 1527-1533). New York, NY: Raven Press.

Schuckit, M.A., Smith, T.L., and Kalmihn, J. (2004). The search for genes contributing to the low level of response to alcohol: Patterns of findings across studies. *Alcoholism: Clinical and Experimental Research, 28*, 1449-1458.

Seidl, R., Peyrl, A., Nicham, R., and Hauser, E. (2000). A taurine and caffeine-containing drink stimulates cognitive performance and well-being. *Amino Acids, 19*, 635-642.

Selden, N.R., Everitt, B.J., Jarrard, L.E., and Robbins, T.W. (1991). Complementary roles for the amygdale and hippocampus in aversive conditioning to explicit and contextual cues. *Neuroscience, 42*, 335-350.

Sharot, T., Martorella, E.A., Delgado, M.R., Phelps, E.A. (2007). How personal experience modulates the neural circuitry of memories of September 11. *Proceedings of the National Academy of Sciences USA, 104*, 389-394.

Sher, L. (2006). Functional magnetic resonance imaging in studies of neurocognitive effects of alcohol use on adolescents and young adults. *International Journal of Adolescent Medicine and Health, 18*, 3-7.

Sher, K.J., Bartholow, B.D., and Nanda, S. (2001). Short- and long-term effects of fraternity and sorority membership on heavy drinking: a social norms perspective. *Psychology of Addictive Behaviors, 15*, 42-51.

Sher, K.J., and Trull, T.J. (1994). Personality and disinhibitory psychopathology: Alcoholism and antisocial personality disorder. *Journal of Abnormal Psychology, 103*, 92-102.

Shih, H.C., Hu, S.C., Yang, C.C., Ko, T.J., Wu, J.K., and Lee, C.H. (2003). Alcohol intoxication increases morbidity in drivers involved in motor vehicle accidents. *American Journal of Emergency Medicine, 21*, 91-94.

Shore, E.R., McCoy, M.L., Toonen, L.A., and Kuntz, E.J. (1988). Arrests of women for driving under the influence. *Journal of Studies on Alcohol, 49*, 7-10.

Shults, R.A., Sleet, D.A., Elder, R.W., Ryan, G.W., and Seghal, M. (2002). Association between state level drinking and driving countermeasures and self reported alcohol impaired driving. *Injury Prevention, 8*, 106-110.

Siegel, S. (1989). Pharmacological conditioning and drug effects. In A.J. Goudie, and M.W. Emmett-Oglesby (Eds.), *Psychoactive Drugs: Tolerance and Sensitization* (pp. 115-181). Clifton, NJ: Humana Press.

Simantov, E., Schoen, C., and Klein, J.D. (2000). Health-compromising behaviors: why do adolescents smoke or drink?: identifying underlying risk and protective factors. *Archives of Pediatric Adolescent Medicine, 154*, 1025-1033.

Simon, M., and Mosher, J. (2007). *Alcohol, Energy Drinks, and Youth: A Dangerous Mix.* San Rafael, CA: Marin Institute.

Sindelar, H.A., Barnett, N.P., and Spirito, A. (2004). Adolescent alcohol use and injury. A summary and critical review of the literature. *Minerva Pediatrica, 56*, 291-309.

Single, E. (1995). Harm reduction and alcohol. *International Journal of Drug Policy, 6*, 26-30.

Sommers, M.S., Dyehouse, J.M., Howe, S.R., Fleming, M., Fargo, J.D., and Schafer, J.C. (2006). Effectiveness of brief interventions after alcohol-related vehicular injury: A randomized controlled trial. *Journal of Trauma, 61*, 523-531.

Spear, L.P., and Varlinskaya, E.I. (2005). Adolescence. Alcohol sensitivity, tolerance, and intake. *Recent Developments in Alcoholism, 17*, 143-159.

Spector, J. (June 26, 2008). A-B to pull caffeine from alcohol drinks. *USA Today,* Retrieved September 11, 2008 from www.usatoday.com.

Spirito, A., Monti, P.M., Barnett, N.P., Colby, S.M., Sindelar, H., Rohsenow, D.J., Lewander, W., and Myers, M. (2004). A randomized clinical trial of brief motivational intervention for alcohol-positive adolescents treated in an emergency department. *Journal of Pediatrics, 145*, 396-402.

Stephens, D.N., Brown, G., Duka, T., and Ripley, T.L. (2001). Impaired fear conditioning but enhanced seizure sensitivity in rats given repeated experience of withdrawal from alcohol. *European Journal of Neuroscience, 14*, 2023-2031.

Stephens, D.N., and Duka, T. (in press). Cognitive and emotional consequences of binge drinking: role of amygdale and prefrontal cortex. *Philosophical Transactions of the Royal Society B.*

Stephens, D.N., Ripley, T.L., Borlikova, G., Schubert, M., Albrecht, D., Hogarth, L., and Duka, T. (2005). Repeated ethanol exposure and withdrawal impairs human fear conditioning and depresses long-term potentiation in rat amygdala and hippocampus. *Biological Psychiatry, 58*, 392-400.

Stockwell, T., McLeod, R., Stevens, M., Philips, M., Webb, M., and Jelinek, G. (2002). Alcohol consumption, setting, gender and activity as predictors of injury: a population-based case-control study. *Journal of Studies on Alcohol, 63*, 372-379.

Straus, R., and Bacon, S.D. (1949). *Drinking in College.* New Haven, CT: Yale University Press.

Substance Abuse and Mental Health Services Administration (SAMHSA) (2007). *Results from the 2006 National Survey on Drug Use and Health: National Findings.* Rockville, MD: SAMHSA Office of Applied Statistics, retrieved October 15, 2008, from http://www.oas.samhsa.gov.

SAMSHA (2005). *Results from the 2004 National Survey on Drug Use and Health: National findings.* Rockville, MD: SAMHSA Office of Applied Statistics.

SAMSHA (2005). National Survey on Drug Use and Health. Office of Applied Studies, Bethesda, MD.

Suomi, S.J. (1999). Behavioral inhibition and impulsive aggressiveness: Insights from studies with rhesus monkeys. In L. Balter, and C.S. Tamis-LeMonda (Eds.), *Child Psychology: a Handbook of Contemporary Issues.* Philadelphia, PA: Taylor and Francis.

Suter, P.M., Schutz, Y., and Yequier, E. (1992). The effect of ethanol on fat storage in healthy subjects. *New England Journal of Medicine, 326*, 695-733.

Swartzwelder, H.S., Richardson, R.C., Markwiese-Foerch, B., Wilson, W.A., and Little, P.J. (1998). Developmental differences in the acquisition of tolerance to ethanol. *Alcohol, 15,* 1-4.

Tabakoff, B., Cornell, N., and Hoffman, P.L. (1986). Alcohol tolerance. *Annals of Emergency Medicine, 15,* 1005-1012.

Tait, R.J., Hulse, G.K., Robertson, S.I., and Sprivulis, P.C. (2002). Multiple hospital presentations by adolescents who use alcohol or other drugs. *Addiction, 97,* 1269-1275.

Tait, R.J., Hulse, G.K., Robertson, S.I., and Sprivulis, P.C. (2005). Emergency department-based intervention with adolescent substance users: 12-month outcomes. *Drug and Alcohol Dependence, 79,* 359-363.

Tjipto, A.C., Taylor, D.M., and Liew, H. (2006). Alcohol use among young adults presenting to the emergency department. *Emergency Medicine Australasia, 18,* 125-130.

Tolstrup, J., Jensen, M.K., Tjanneland, A., Overvad, K., Mukamel, M.J., and Granbaak, M. (2006). Prospective study of alcohol drinking and coronary heart disease in women and men. *British Medical Journal, 332,* 1244-1248.

Tormey, W.P., and Bruzi, A. (2001). Acute psychosis due to the interaction of legal compounds: Ephedra alkaloids in "Vigueur Fit" tablets, caffeine in "Red Bull", and alcohol. *Medical Science Law, 41,* 331-336.

Townshend, J.M., and Duka, T. (2001). Attentional bias associated with alcohol cues: differences between heavy and occasional social drinkers. *Psychopharmacology, 157,* 67-74.

Townshend, J.M., and Duka, T. (2005). Binge drinking, cognitive performance and mood in a population of young social drinkers. *Alcoholism: Clinical and Experimental Research, 29,* 317-325.

Treutlein, J., Kissling, C., Frank, J., Wiemann, S., Dong, L., Depner, M., Saam, C., Lascorz, J., Soyka, M., Preuss, U.W., Rujescu, D., Skowronek, M.H., Rietschel, M., Spanagel, R., Heinz, A., Laucht, M., Mann, K., and Schumann, G. (2006). Genetic association of the human corticotrophin releasing hormone receptor 1 (CRHR1) with binge drinking and alcohol intake patterns in two independent samples. *Molecular Psychiatry, 11,* 594-602.

Trull, T.J., Waudby, C.J., and Sher, K.J. (2004). Alcohol, tobacco, and drug use disorders and personality disorder symptoms. *Experimental and Clinical Psychopharmacology, 12,* 65-75.

Tucker, J.S., Ellickson, P.L., and Klein, D.J. (2008). Growing up in a permissive household: what deters at-risk adolescents from heavy drinking? *Journal of Studies on Alcohol and Drugs, 69,* 528-534.

Turner, J.C., and Shu, J. (2004). Serious health consequences associated with alcohol use among college students: demographic and clinical characteristics of patients seen in an emergency department. *Journal of Studies on Alcohol, 65,* 179-183.

U.S. Department of Health and Human Services (2000). *Healthy People 2010: Understanding and Improving Health and Objectives for Improving Health* (2nd ed., Vol. 2, Goal 26-11:

Reduce the proportion of persons engaging in binge drinking of alcoholic beverages). Washington, DC: U.S. Government Printing Office.

U.S. Department of Health and Human Services (2007). *The Surgeon General's Call to Action to Prevent and Reduce Underage Drinking.* U.S. Department of Health and Human Services, Office of the Surgeon General.

US Preventive Services Task Force (1996). *Guide to Clinical Preventive Services* (2nd ed.). Baltimore, MD: Williams and Wilkins.

Van Eden, C.G., Kros, J.M., and Uylings, H.B.M. (1990). The development of the rat prefrontal cortex: Its size and development of connections with thalamus, spinal cord and other cortical areas. In H.B.M. Uylings, C.G. van Eden, J.P.C. De Bruin, M.A. Corder and M.G.P. Feenstra (Eds.), *Progress in Brain Research: Vol. 85. The Prefrontal Cortex: Its Structure, Function and Pathology* (pp. 169-183). Amsterdam: Elsevier Science.

Viner, R.M., and Taylor, B. (2007). Adult outcomes of binge drinking in adolescence: findings from a UK national birth cohort. *Journal of Epidemiology and Community Health, 61,* 902-907.

Vinson, D.C., Maclure, M., Reidinger, C., and Smith, G.S. (2003). A population-based case-crossover and case-control study of alcohol and the risk of injury. *Journal of Studies on Alcohol, 64,* 358-366.

Voas, R.B., Lange, J.E., and Johnson, M.B. (2002). Reducing high-risk drinking by young Americans south of the border: the impact of a partial ban on sales of alcohol. *Journal of Studies on Alcohol, 63,* 286-292.

Voas, R.B., Romano, E., Kelley-Baker, T., and Tippetts, A.S. (2006). A partial ban on sales to reduce high-risk drinking South of the border: seven years later. *Journal of Studies on Alcohol, 67,* 746-753.

Voas, R.B., Tippetts, A.S., and Fell, J.C. (2003). Assessing the effectiveness of minimum legal drinking age and zero tolerance laws in the United States. *Accident Analysis and Prevention, 35,* 579-587.

Vogel-Sprott, M. (1992). *Alcohol tolerance and social drinking: Learning the consequences.* New York: Guilford.

Vogel-Sprott, M., and Fillmore, M.T. (1999). Expectancy and behavioral effects of socially used drugs. In I. Kirsch (Ed.), *How Expectancies Shape Experience* (pp. 215-232). Washington, DC: American Psychological Association.

Volkow, N.D., Fowler, J.S., and Wang, G.J. (2003). The addicted human brain: insights from imaging studies. *Journal of Clinical Investigation, 111,* 1444-1451.

Volkow, N.D., Fowler, J.S., and Wang, G.J. (2004). The addicted human brain viewed in the light of imaging studies: brain circuits and treatment strategies. *Neuropharmacology, 47 (Suppl. 1),* 3-13.

von Diemen, L., Bassani, D.G., Fuchs, S.C., Szobot, C.M., and Pechansky, F. (2008). Impulsivity, age of first alcohol use and substance use disorders among male adolescents: a population based case-control study. *Addiction, 103,* 1198-1205.

Wagenaar, A.C., O'Malley, P.M., and LaFond, C. (2001). Lowered legal blood alcohol limits for young drivers: Effects on drinking, driving, and driving-after-drinking behaviors in 30 states. *American Journal of Public Health, 91,* 801-804.

Wallace, A.E., Wallace, A., and Weeks, W.B. (2008). The U.S. military as a natural experiment: changes in drinking age, military environment, and later alcohol treatment episodes among veterans. *Military Medicine, 173,* 619-625.

Waller, P.F., Hill, E.M., Maio, R.F., and Blow, F.C. (2003). Alcohol effects on motor vehicle crash injury. *Alcoholism: Clinical and Experimental Research, 27,* 695-703.

Walters, S.T., Bennett, M.E., and Noto, J.V. (2000). Drinking on campus. What do we know about reducing alcohol use among college students? *Journal of Substance Abuse Treatment, 19,* 223-228.

Warburton, D.M., Bersellini, E., and Sweeney, E. (2001). An evaluation of a caffeinated taurine drink on mood, memory and information processing in healthy volunteers without caffeine abstinence. *Psychopharmacology, 158,* 322-328.

Watt, K., Purdie, D.M., Roche, A.M., and McClure, R.J. (2004). Risk of injury from acute alcohol consumption and the influence of cofounders. *Addiction, 99,* 1262-1273.

Watt, K., Purdie, D.M., Roche, A.M., and McClure, R.J. (2005). The relationship between acute alcohol consumption and consequent injury type. *Alcohol and Alcoholism, 40,* 263-268.

Weafer, J., and Fillmore, M.T. (2008). Alcohol impairment of inhibitory mechanisms in binge and non-binge drinkers: Examining the role of inhibitory control in alcohol abuse. Poster presentation at the 31st Annual Meeting of the Research Society on Alcoholism, Chicago, IL.

Weafer, J., and Fillmore, M.T. (2008). Individual differences in acute alcohol impairment of inhibitory control predict ad libitum alcohol consumption. *Psychopharmacology, 201,* 315-324.

Wechsler, H., and Austin, S.B. (1998). Binge drinking: The five/four measure [Letter to the editor]. *Journal of Studies on Alcohol, 59,* 122-123.

Wechsler, H., Davenport, A., Dowdall, G.W., Moeykens, B., and Castillo, S. (1994). Health and behavioral consequences of binge-drinking in college: A national survey of students at 140 campuses. *Journal of the American Medical Association, 272,* 1672-1677.

Wechsler, H., Dowdall, G.W., Davenport, A., and Castillo, S. (1995). Correlates of college student binge drinking. *American Journal of Public Health, 85,* 921-926.

Wechsler, H., Dowdall, G.W., Davenport, A., and Rimm, E.B. (1995). A gender-specific measure of binge drinking among college students. *American Journal of Public Health, 85,* 982-985.

Wechsler, H., Dowdall, G.W., Maenner, G., Geldhill-Hoyt, J., and Lee, H. (1998). Changes in binge drinking and related problems among American college students between 1993 and 1997. *Journal of American College Health, 47,* 57-68.

Wechsler, H., Kelley, K., Weitzman, E.R., SanGiovanni, J.P., and Seibring, M. (2000). What colleges are doing about student binge drinking. A survey of college administrators. *Journal of American College Health, 48,* 219-226.

Wechsler, H., and Kuo, M. (2000). College students define binge drinking and estimate its prevalence: results of a national survey. *Journal of American College Health, 49,* 57-64.

Wechsler, H., Lee, J.E., Kuo, M., and Lee, H. (2000). College binge drinking in the 1990s: A continuing problem. *Journal of American College Health, 48,* 199-210.

Wechsler, H., Lee, J.E., Kuo, M., Seibring, M., Nelson, T.F., and Lee, H.P. (2002). Trends in college binge drinking during a period of increased prevention efforts: Findings from four Harvard School of Public Health study surveys, 1993-2001. *Journal of American College Health, 50,* 203-217.

Wechsler, H., Lee, J.E., Nelson, T.F., and Lee, H. (2003). Drinking and driving among college students: The influence of alcohol-control policies. *American Journal of Preventative Medicine, 25,* 212-218.

Wechsler, H., Moeykens, B., Davenport, A., Castillo, S., and Hansen, J. (1995). The adverse impact of heavy episodic drinkers on other college students. *Journal of Studies on Alcohol, 56,* 628-634.

Wechsler, H., and Nelson, T.F. (2001). Binge drinking and the American college students: What's five drinks? *Psychology of Addictive Behaviors, 15,* 287-291.

Wechsler, H., and Nelson, T.F. (2006). Relationship between level of consumption and harms in assessing drink cut-points for alcohol research: Commentary on "Many college freshman drink at levels far beyond the binge threshold" by White et al. *Alcoholism: Clinical and Experimental Research, 30,* 922-927.

Wechsler, H., Nelson, T.F., Lee, J.E., Seibring, M., Lewis, C., and Keeling, R.P. (2003). Perception and reality: A national evaluation of social norms marketing interventions to reduce college students' heavy alcohol use. *Journal of Studies on Alcohol, 64,* 484-494.

Wechsler, H., Seibring, M., Liu, I.C., and Ahl, M. (2004). Colleges respond to student binge drinking: reducing student demand or limiting access. *Journal of American College Health, 52,* 159-168.

Wechsler, H., and Wuethrich, B. (2002). *Dying to Drink: Confronting Binge Drinking on College Campuses.* U.S.A., Rodale.

Weinberg, L., and Wyatt, J.P. (2006). Children presenting to hospital with acute alcohol intoxication. *Emergency Medicine Journal, 23,* 774-776.

Weiss, M., Hechtman, L.T., and Weiss, G. (1999). *ADHD in adulthood: A guide to current theory, diagnosis and treatment.* Baltimore, MA: Johns Hopkins University Press.

Weissenborn, R., and Duka, T. (2003). Acute alcohol effects on cognitive function in social drinkers: their relationship to drinking habits. *Psychopharmacology, 165,* 306-312.

Weitzman, E.R., Nelson, T.F., and Wechsler, H. (2003). Taking up binge drinking in college: the influences of person, social group, and environment. *Journal of Adolescent Health, 32,* 26-35.

Wells, S., Mihic, L., Tremblay, P.F., Graham, K., and Demers, A. (2008). Where, with whom, and how much alcohol is consumed on drinking events involving aggression? Event-level associations in a Canadian national survey of university students. *Alcoholism: Clinical and Experimental Research, 32,* 522-533.

Wender, P.H. (1995). *Attention-deficit hyperactivity disorder in adults.* New York, NY: Oxford University Press.

West, C. (2007). Why it's impossible for some to 'Just Say No'. *Observer, 20(10),* 8.

West, S.L., and O'Neal, K.K. (2004). Project D.A.R.E. outcome effectiveness revisited. *American Journal of Public Health, 94,* 1027-1029.

White, A.M., Kraus, C.L., Flom, J.D., Kestenbaum, L.A., Mitchell, J.R., Shah, K., and Swartzwelder, H.C. (2005). College students lack knowledge of standard drink volumes: Implications for definitions of risky drinking based on survey data. *Alcoholism: Clinical and Experimental Research, 29,* 631-638.

White, A.M., Kraus, C.L., and Swartzwelder, H.S. (2006). Many college freshman drink at levels far beyond the binge threshold. *Alcoholism: Clinical and Experimental Research, 30,* 1006-1010.

White, A.M., Kraus, C.L., McCracken, L.A., and Swartzwelder, H.S. (2003). Do college students drink more than they think? Use of a free-pour paradigm to assess how college students define standard drinks. *Alcoholism: Clinical and Experimental Research, 24,* 1751-1756.

White, H.R., Mun, E.Y., and Morgan, T.J. (2008). Do brief personalized feedback interventions work for mandated students or is it just getting caught that works? *Psychology of Addictive Behaviors, 22,* 107-116.

Wilens, T.E., Faraone, S.V., Biederman, J., and Gunawardene, S. (2003). Does stimulant therapy of attention-deficit/hyperactivity disorder beget later substance abuse? A meta-analytic review of the literature. *Pediatrics, 111,* 179-185.

Williams, A.F., Lund, A.K., Preusser, D.F. (1986). Drinking and driving among high school students. *The International Journal of the Addictions, 21,* 643-655.

WKRC TV Cincinnati (2008). Miami students pleads guilty in alcohol death. *WKRC TV Cincinnati, OH, January 24, 2008.*

Wolfer, D.P., and Lipp, H.-P. (1995). Evidence for physiological growth of hippocampal mossy fiber collaterals in the guinea pig during puberty and adulthood. *Hippocampus, 5,* 329-340.

Woolfenden, S., Dossetor, D., and Williams, K. (2002). Children and adolescents with acute alcohol intoxication/self-poisoning presenting to the emergency department. *Archives of Pediatric and Adolescent Medicine, 156,* 345-348.

World Health Organziation (WHO) (2001, February). *Declaration on Young People and Alcohol.* Paper presented at the World Health Organization European Ministerial Conference on Young People and Alcohol, Stockholm.

WHO (2001). *Global Status Report: Alcohol and Young People.* Geneva, Switzerland: Author. Retrieved September 25, 2008, from http://www.who.int.

WHO (2004). *Global Status Report: Alcohol Policy.* Geneva, Switzerland: Author. Retrieved September 25, 2008, from http://www.who.int.

WHO (2007). *Alcohol and Injury in Emergency Departments: Summary of the Report from the WHO Collaborative Study on Alcohol and Injuries.* Geneva, Switzerland: Author. Retrieved October 13, 2008, from http://www.who.int.

Wright, S.W., and Clovis, C.M. (1996). Drinking on campus. Undergraduate intoxication requiring emergency care. *Archives of Pediatric and Adolescent Medicine, 150,* 699-702.

Wright, S.W., Norton, V.C., Dake, A.D., Pinkston, J.R., and Slovis, C.M. (1998). Alcohol on campus: alcohol-related emergencies in undergraduate college students. *Southern Medical Journal, 91,* 909-913.

Wright, S.W., and Slovis, C.M. (1996). Drinking on campus: undergraduate intoxication requiring emergency care. *Archives of Pediatric and Adolescent Medicine, 150,* 699-702.

Xing, Y., Ji, C., and Zhang, L. (2006). Relationship of binge drinking and other health-compromising behaviors among urban adolescents in China. *Journal of Adolescent Health, 39,* 495-500.

Young, S.Y., Hansen, C.J., Gibson, R.L., and Ryan, M.A. (2006). Risky alcohol use, age at onset of drinking, and adverse childhood experiences in young men entering the US Marine Corp. *Archives of Pediatric Adolescent Medicine, 160,* 1207-1214.

Zador, P.L., Krawchuk, S.A., and Voas, R.B. (2000). Alcohol-related relative risk of driver fatalities and driver involvement in fatal crashes in relation to driver age and gender: an update using 1996 data. *Journal of Studies on Alcohol, 61,* 387-395.

Zakrajsek, J.S., and Shope, J.T. (2006). Longitudinal examination of underage drinking and subsequent drinking and risky driving. *Journal of Safety Research, 37*, 443-451.

Zhang, L., and Johnson, W.D. (2005). Violence-related behaviors on school property among Mississippi Public High School students, 1993-2003. *Journal of School Health, 75*, 67-71.

Ziegler, D.W., Wang, C.C., Yoast, R.A., Dickinson, B.D., McCaffreee, M.A., Robinowitz, C.B., Sterling, M.L., Council on Scientific Affairs, American Medical Association (2005). The neurocognitive effects of alcohol on adolescents and college students. *Preventative Medicine, 40*, 23-32.

Zuckerman, M. (1979). *Sensation seeking: Beyond the optimal level of arousal.* New York, NY: John Wiley.

Zuckerman, M. (1984). Sensation seeking: a comparative approach to a human trait. *Behavioral and Brain Sciences, 7*, 413-471.

13 Wham-TV (2005). Taverns shy away from liquid charge. Retrieved December 6, 2005, from www.13wham.com.

INDEX

J

K

S

Y

Z